ENERGY AND INFRASTRUCTURE

Volume 5

The Environmental Management of Low-Grade Fuels

Full list of titles in the set
ENERGY AND INFRASTRUCTURE

The Environmental Management of Low-Grade Fuels

Mary MacDonald, Michael Chadwick and Gareg Aslanian

from Routledge

First published by Earthscan in the UK and USA in 1996

For a full list of publications please contact:
Earthscan
2 Park Square, Milton Park, Abingdon, Oxfordshire OX14 4RN
711 Third Avenue, New York, NY 10017

First issued in paperback 2016

Earthscan is an imprint of the Taylor & Francis Group, an informa business

ISBN 13: 978-1-138-98916-0 (pbk)
ISBN 13: 978-1-84407-977-3 (hbk)

ISBN 978-1-84407-972-8 (Energy and Infrastructure set)
ISBN 978-1-84407-930-8 (Earthscan Library Collection)

Earthscan publishes in association with the International Institute for Environment and Development

A catalogue record for this book is available from the British Library

Library of Congress Cataloging-in-Publication Data has been applied for

Publisher's note
The publisher has made every effort to ensure the quality of this reprint, but points out that some imperfections in the original copies may be apparent.

At Earthscan we strive to minimize our environmental impacts and carbon footprint through reducing waste, recycling and offsetting our CO_2 emissions, including those created through publication of this book. For more details of our environmental policy, see www.earthscan.co.uk.

The
Environmental Management
of
Low-Grade Fuels

A Study Prepared for the United Nations Environment Programme by the Stockholm Environment Institute with the Assistance of the Institute for High Temperatures and the Centre for International Projects, Moscow

The Environmental Management of Low-Grade Fuels

M.E. MacDonald
M.J. Chadwick
G.S. Aslanian

With contributions from
B.S. Beloselsky, L. Kropp, V.M. Maslenikov and M. Prior

STOCKHOLM ENVIRONMENT INSTITUTE
STOCKHOLM
1996

EARTHSCAN
Earthscan Publications Ltd, London

First published in the UK in 1996 by
Earthscan Publications Limited

Copyright © Stockholm Environment Institute 1996

A catalogue record for this book is available from the British Library

ISBN 1 85383 339 8 Hardback

Cover design by Erik Willis and Steve Cinderby

For a full list of publications please contact:

**Earthscan Publications Limited
120 Pentonville Road
London N1 9JN
Tel: 0171 278 0433
Fax: 0171 278 1142**

Earthscan is an editorially independent subsidiary of Kogan
Page Limited and publishes in association with WWF-UK and
the International Institute for Environment and Development

Contents

List of Figures and Tables

Preface

The information presented here focuses on environmental protection during the use of low-grade fuels such as peat, wood, biomass, lignite, oil shale, and municipal and industrial wastes. Specific technical strategies and policy options are given for reducing or eliminating the negative impacts associated with the mining, transport, processing and combustion of these fuels. Throughout the book, discussions of environmental protection measures are summarized in table form for easy reference. It is hoped that the chapters which follow will prove helpful to industry and government personnel responsible for ensuring the sound environmental management of low-grade sources of energy. An attempt has also been made to present the material in a straightforward manner so that anyone wishing to learn more about low-grade fuels will find it useful.

As we approach the twenty-first century, environmental issues surrounding the production and use of energy continue to be problematic and complex. Negative environmental impacts arising from energy use include pollution of air, ground and surface waters and soil due to the production and release of emissions and effluents and the build-up of solid waste throughout the fuel cycle. A decline in the quality of the natural environment will have a disruptive or destructive effect on vegetation, fish and wildlife habitat, existing and future land use and human health. In addition, fuel combustion contributes to atmospheric levels of carbon dioxide, ozone, methane and nitrogen oxide, gases which make a major contribution to global warming.

Economic growth through industrial development using high calorific value fuels remains an important objective of many nations. The gap between countries which can afford to ensure a dependable supply of energy to meet their energy needs and those which cannot continues to grow, particularly with respect to high quality fossil fuels such as oil, natural gas and hard coal.

Ultimately, we must work towards a world where the efficient use of energy, and its conservation, is a given and renewable energy, managed in a sustainable and environmentally responsible manner, provides the majority of energy for all sectors of the economy in all parts of the world. When making decisions or developing energy-related policy it is important to bear in mind that in the long run renewable energy sources are the only environmentally satisfactory option.

However, the reality is that many nations will choose to pursue economic development using non-renewable fuels or poorly managed biomass for as long as possible. While current costs and uncertainty surrounding renewable forms of energy persist in the short term, their widespread application to meet the energy needs of the industrial and transportation sectors of most countries seems unlikely. If high quality fossil fuels are unavailable, the use of low-grade alternatives such as peat, lignite, oil shale and wood may increase. For many nations, the wise environmental management of these fuels may, from necessity, prove to be an important link to a more environmentally sound energy future.

Acknowledgements

The authors wish to acknowledge the United Nations Environment Programme (UNEP) and the Stockholm Environment Institute (SEI) for their sponsorship of the work on the environmental management of low-grade fuels presented in this volume. We appreciate the interest and support given by the Centre for International Projects (CIP), representing UNEP in Moscow. In addition, we would like to express our gratitude to a number of individuals including Professor Gordon Goodman, former director of SEI, Dr Vladimir Simonov, formerly at CIP and now at IVTAN in Moscow, and Dr John Christensen at the UNEP Collaborating Centre on Energy and Environment, Risø National Laboratory, Denmark, who were instrumental in the initiation of the project and who gave continued encouragement throughout. Dr Igor Radomsky and Dr Mikail Novikov at CIP provided technical and logistical assistance. From Mr Peter Bailey (SEI) and Dr Everett Sondreal (University of North Dakota) we received valuable information for inclusion in the work.

A full review of the material presented here was conducted by Dr Leonard Hamilton of the Biomedical and Environmental Assessment Division of the National Center for the Analysis of Energy Systems at the Brookhaven National Laboratory, Upton, NY, USA, and by Mr Jussi-Pekka Aittola of Viatak, Espoo, Finland. We are indebted to both reviewers for their informed and comprehensive comments.

Thanks are also due to Mrs Sue Sparrow for the graphics accompanying the text, to Ms Rowan Davies for editorial advice, to Dr Arno Rosemarin for production assistance, to Erik Willis and Steve Cinderby for the creative artwork that they produced for the cover of this volume, and to Mrs Isobel Devane for her careful editing, word processing and general support in the preparation of this work.

Finally, we are grateful to the following individuals and organizations who have kindly given permission for the reproduction of copyright material:

Australasian Institute of Mining and Metallurgy (Figure 4.1); Blackwell Scientific Publications (Figure 2.2); CRC Press, Florida (Figure 3.3); Elsevier Applied Science Publishers (Figure 3.2); Financial Times Business Information, London (Figure 7.1); *Fuel* (Figure 2.1; Table 3.3); Götaverken Energy AB, Stockholm (Figure 7.7); Dr J R Howard, Solihull (Figure 7.3);

Dr H Huettenhain, San Francisco (Table 6.2); IEA Coal Research, London (Figure 6.2, Figure 6.3 and Figure 7.18; Table 6.3); IHI, Tokyo (Figure 7.2); International Development Research Centre, Ottawa (Table 3.6); IOP Publishing, Bristol (Figure 7.3); Minnesota Department of Natural Resources (Table 1.1); New York State Energy Research and Development Authority, Albany (Table 4.9); Professor Adilson de Oliveira, Rio de Janeiro (Figure 9.1); Pergamon Press, Oxford (Figure 1.2); Professor R F Prcbstein, Cambridge, MA (Figure 6.6); Dr K Riedle and Dr B Böhm, Erlangen (Figure 8.1); *Science and Public Policy* (Figure 9.1); Shell International Petroleum Company, London (Figure 8.2); Professor A J Smith, Surrey (Figure 3.1); State Electricity Commission of Victoria, Melbourne (Figure 1.3); Technical Research Centre of Finland (Table 5.1 and Table 7.1; Figure 7.5 and Figure 7.6); The World Bank (Table 3.2); UNEP, Nairobi (Figure 6.5); United Engineers and Constructors/Philadelphia Electric Company (Figure 7.14); US Congress, Office of Technology Assessment (Figure 1.7); Valgus, Tallinn (Figure 2.3); Verlag Glückauf GmbH (Table 1.8); VGB Kraftwerkstechnik, Essen (Table 8.2); World Energy Conference (Table 3.1, Table 3.4, Table 3.5 and Table 3.7).

Foreword

'Energy is dangerous', or so asserts the preface of a recent book on energy alternatives. 'Energy is good for you!' Or so it might seem if we look at the positive relationships between per capita energy use and life expectancy or the negative relationship between per capita use and infant mortality. We should probably have to agree that 'energy is complicated' – particularly in the relationships between energy supply and demand on the one hand and development, welfare and environmental considerations on the other. And if this holds for the more conventional energy sources – oil, gas, coal, even nuclear – then considerations affecting the use and impacts of low-grade fuels are even more complicated. 'Low-grade fuel' implies a low calorific value per unit weight. But the result of this may be a higher percentage composition of potentially polluting material emitted under certain methods of fuel utilization. Technologies exist to reduce or almost completely eliminate polluting emissions or effluents, and to handle the solid wastes in an environmentally benign way, but these are often relatively costly – and a percentage of such costs are usually incurred continuously.

Many countries, increasing in number, seek apparently low-cost fuels of indigenous origin to meet their energy security arrangements. Consequently, materials that are of high potential value as a chemical feedstock, or may find use for other purposes but are low-grade fuels, are combusted or otherwise converted as an energy source.

The United Nations Environment Programme and the Stockholm Environment Institute have well-documented histories of concern for the efficient and rational use of energy in a way that fully accepts the requirement to protect the environment and reduce as far as possible risks to human health. The study that forms the basis of this volume was undertaken to continue that concern into the area of the fuels which are less generally given consideration from this point of view.

The purpose of the volume is to introduce the background and range of considerations that energy planners require if efficient and rational energy use via low-grade fuels is to be combined with environmental protection concerns.

E. Dowdeswell
Executive Director
United Nations Environment
 Programme
Nairobi

M.J. Chadwick
Director
Stockholm Environment Institute
Stockholm

Overview

The emphasis here is placed on the environmental impacts associated with the use of low-grade solid fuels along with the identification of measures which will mitigate the potential negative effects of mining or harvesting, transporting, processing and using low-grade solid fuels. The concept of environment is used in its broadest sense and is meant to include the physical, biological, health, social and economic components of the environment as well as the interaction between and among these individual components.

A definition of low-grade fuels

The term 'low-grade' when applied to fuel has assumed a number of meanings in the literature. Low-grade has been used to describe fuels with a high percentage of potential pollutants such as fuels with a high ash or high sulphur content (Beloselsky, Solyakov and Pokrovsky, 1992; Hamilton, 1992; IDRC, 1986). Fuels with a high moisture content have also been labelled as low-grade (BCRA, 1987; IEA, 1983). Other definitions of low-grade fuels have focused on the calorific value of a fuel in the 'as fired' stage of fuel utilization. There are also secondary products of fuel production or utilization (e.g. biogas) which have been referred to as low grade (LIEE, 1994; Kjellström, 1985).

Low-grade fuels have also been known as low-quality or poor-quality fuels or as fuels containing a combination of characteristics such as high ash, low carbon, high moisture, high mineral and high sulphur (Beloselsky, Solyakov and Pokrovsky, 1992; Aittola, 1992a; Chadwick, 1992; Hyöty, 1992; ECE/UNDP, 1986; IDRC, 1986).

Traditionally, the term 'grade' has been used to refer to the potential end use of a given coal but has increasingly been applied to all fuels. The most common end use is combustion, with other processes such as gasification, pyrolysis and liquefaction being considered as accompanying stages of fuel combustion. The heat given off during combustion of a standard quantity of fuel at a standard temperature is known as its calorific value.

The amount of carbon available for conversion to heat is a primary determinant of the calorific value of a fuel. Components such as ash, moisture, oxygen and sulphur are referred to as 'impurities' and, in general, the greater the quantity of impurities contained in a fuel the lower its calorific value.

The highest calorific value solid fuel is anthracite, which has a calorific value of approximately 35 megajoules per kilogram (8365 kcal/kg) for the best-quality hard coal. For purposes of this study, it was determined that half of this value would be a reasonable and workable dividing line between high and low calorific fuels (Figure 1.1).

The fuels referred to in this study are considered low grade if they have a calorific value of roughly less than 17.5 MJ/kg[1] (4182 kcal/kg). This is the primary characteristic that is used. A calorific value greater than 17.5 MJ/kg and an impurity content with a potentially significant environmental impact such as a sulphur content greater than 3 per cent or an ash content greater than 20 per cent is a secondary criterion.

Prior to processing in an 'as received' state, solid fuels will be found in a broad range of calorific values. Although certain locales and situations may yield fuel with a higher calorific value (e.g. different types of municipal or industrial waste), those which generally fall below 17.5 MJ/kg include:

- wood;
- biomass other than wood (including agricultural waste);
- municipal waste;
- industrial waste;
- mining waste (e.g. tailings and fines);
- peat;
- oil shale; lignite/brown coal;
- some sub-bituminous coals.

Solid fuels which have a calorific value greater than 17.5 MJ/kg but which may have an impurity content which renders them low grade include a small percentage of:

- sub-bituminous coal;
- bituminous coal;
- anthracite.

These fuels are not a primary focus of the work presented here due to a lack of data regarding the impurity content of higher-rank coals. However, they are included in the discussion where information is available and their inclusion is relevant.

---

1 The figure of 35 MJ/kg may also be read as gigajoules per tonne (GJ/t) since the proportions are the same. Kilocalories per kilogram (kcal/kg) may be roughly calculated from MJ by dividing by 4.184 and multiplying by 1000.

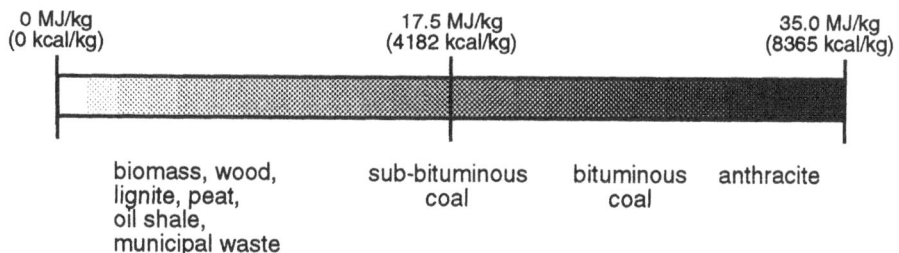

Figure 1.1 *Relative calorific value of solid fuels – 'as received'*

Factors contributing to calorific value

The quantity of combustible carbon contained in a fuel is a major contributing factor to its calorific value. The presence of oxygen in a fuel is also important, since oxygen reduces the amount of hydrogen that will be used during combustion. Figure 1.2 illustrates the relationship between hydrogen and oxygen in a number of solid fuels.

Volatile matter – that is, matter which evaporates during the combustion of a fuel – may or may not contribute to its calorific value. The volatile matter of coal, for instance, can vary widely (Couch, 1988). High-quality anthracite has virtually no volatile matter, while some high calorific bituminous coals have the same percentage as lignite (Harker and Backhurst, 1981).

The moisture content of a solid fuel can also affect its heat value (Couch, 1988; Boron, Evans and Peterson, 1987). A higher moisture content generally indicates that there is less carbon available for conversion, and therefore fuels containing a large percentage of moisture tend to have low calorific values in an untreated state.

Environmental implications of calorific value

Fuels which have a low calorific value prior to any treatment or processing may, ultimately, be upgraded to high calorific value fuels through the application of a variety of beneficiation procedures and technologies. Certain peats, for instance, simply require drying to increase the heat produced during combustion. Oil shales require a more complicated, expensive and potentially polluting series of pre-combustion or combustion treatments to increase calorific value.

The definition of low-grade fuels used here is based primarily on the calorific value of the fuel. The importance of this aspect of the definition lies in the recognition that using fuels of lower calorific value to replace

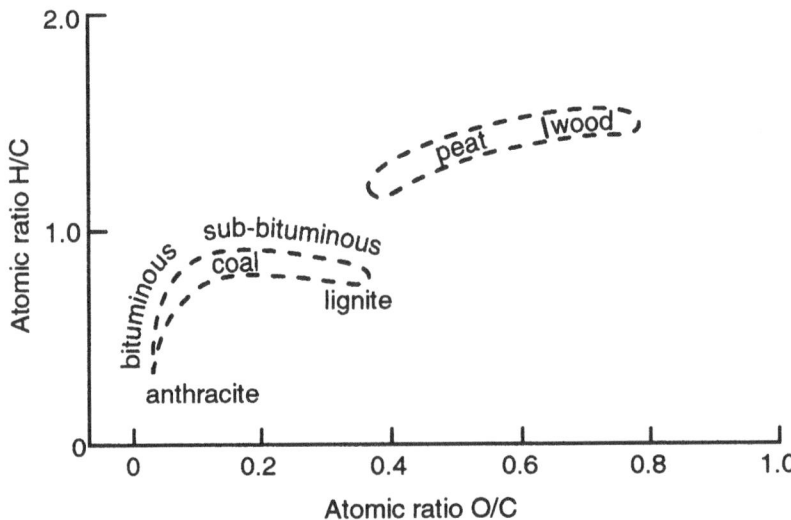

Figure 1.2 *Relative atomic compositions of several fuel types*

(Source: Chadwick, Highton and Lindman, 1987)

higher calorific value fuels poses a number of potential technological and environmental problems.

If the fuels are used 'as received' or untreated, the combustion of large amounts of lower grade fuels will be needed to achieve an energy output equal to that of a smaller amount of higher grade fuel. This may result in the release, pro rata, of large quantities of emissions such as carbon dioxide (CO_2), water vapour, sulphur dioxide (SO_2), nitrogen oxides (NO_x) and ash into the environment.

Processing of low-grade fuel can produce a cleaner burning fuel. However, while reducing the potential for environmental degradation during combustion, upgrading will still result in contaminated effluent and process gas and a build-up of solid wastes and residues. These problems are not eliminated; they simply occur at an earlier stage of the fuel cycle and must be dealt with there.

Low-grade fuels considered

The issues and environmental impacts associated with the use of low-grade solid fuels will be illustrated using the examples of peat, lignite, oil shale and wood. Information pertaining to biomass other than wood, such as agricultural waste, to municipal, mining and industrial waste and to high-impurity high calorific value coals has been included. It must be pointed out that wastes from different sources and processes are notoriously variable in their properties and characteristics and often cannot be readily placed in any overall classification system or scheme.

Characteristics of low-grade fuels

Peat

Peat is a low-grade fuel since, when it undergoes combustion with the maximum amount of moisture with which burning can occur, it will have a calorific value of only 4–8.2 MJ/kg (956–1960 kcal/kg) (Aittola, 1992a; Monenco, 1981). Prior to drying or processing, many newly harvested peats would not be suitable for use as a fuel due to a very high moisture content.

Peat, the youngest of the fossil fuels, is formed during a build-up of plant material under waterlogged conditions when the accumulation of vegetation occurs more quickly than its decay. Recorded rates of accumulation indicate an average of 1.66 mm per year (WEC, 1989). Layers of plant material are laid down over time, which gives peat a stratified structure.

The amount of organic material in peat varies greatly, from 20 per cent to 99 per cent (Boron, Evans and Peterson, 1987; Monenco, 1981). Some may comprise distinguishable plant remains but is mainly very fine, well-decomposed organic matter. For geological purposes 30 cm of accumulated decomposed plant matter is called peat (Lappalainen, 1992), but only a 1-m accumulation is recognized as sufficient for mining purposes (Boron, Evans and Peterson, 1987).

Types of peat A great deal of work has been done with the aim of developing a comprehensive system for classifying peats. One reason for this is that the properties of peat can vary, to a large degree, from one location to another, and the end use of peat is often determined by the specific characteristics of an individual deposit. The von Post method for classifying peat has been in common usage since the early 1920s. With this system peat is assigned to one of ten possible categories (H1–H10) following the squeezing of a sample of peat in the hand. The degree of decomposition of plant material and the moisture level are the primary factors influencing assignment to a particular class. The International Peat Society has reduced these ten categories down to three general classifications including peat moss (fibric peat), reed-sedge (hemic peat) and humus (sapric peat). A summary of the characteristics of these three classes of peat is given in Table 1.1.

Table 1.1 *Characteristics of three major categories of peat* [a]

Category	Per cent of fibre	Structure and appearance of the peat	Presence and appearance of humus
Fibric (peat moss)	>70% weakly decomposed peat	Spongy or fibrous, built of plant residues tied with one another; for separation, tearing of plant residues is required; easily recognizable plant residues (well preserved); elastic, compact	Not visible or occurs in small amounts as a dispersed dark mass, saturating and colouring plant residues
Hemic peat	40–70% medium decomposed peat	Amorphous-fibrous, most contain numerous plant residues of various sizes; wood peats are more friable due to the presence of wood residues in amorphous humus; when pressed in fingers, transforms into an amorphous, plastic mass	Distinctly discernible against which plant residues are visible; humus can be pressed out between fingers of the clenched fist but not more than one-third of the taken sample
Sapric peat	<40% strongly decomposed peat	Lumpy-amorphous, consisting in main part of humus; in lumpy amorphous peat greater fragments of plant residue/wood, rhizomes, greater rootlets occur; friable, disintegrates under pressure; amorphous peat strongly plastic, with sporadic large fragments of plant residues	Uniform mass can be pressed out between fingers of the clenched fist in the amount of a half or the whole of the taken sample

Amount and appearance of water	Potential use as a fuel	Elemental composition (% of dry organic material)[d]		
Great amount of water, which can be easily pressed out and pours as a streamlet; almost totally pure or slightly brownish; may contain dark humus spots	Poor quality fuel due to high water retention capacity of the weakly decomposed plant material[b] ; ash content approximately 1.25%	C	48.0–53.0	
		H	5.0–6.10	
		O_2	40.0–46.0	
		N	0.50–1.00	
		S	0.10–0.20	
Can be pressed out or flows by few drops; usually thick and of dark colour/humus; in drained peat, slightly coloured with humus coagulated in consequence of partial drying	Good quality fuel due to high level of fixed carbon and low ash content[b]; ash content range 4–7%[e, f]	C	56.0–58.0	
		H	5.50–6.10	
		O	34.0–39.0	
		N	0.80–1.10	
		S	0.10–0.30	
Cannot be pressed out, instead the humus mass is squeezed	Poor quality fuel due to high moisture content and ash levels as high as 60%[c] ; (high ash sapric peat known as muck)	C	59.0–63.0	
		H	5.10–6.10	
		O	31.0–34.0	
		N	0.90–1.90	
		S	0.20–0.50	

[a] Modified slightly from Minnesota Department of Natural Resources (1984) except where indicated
[b] Boron, Evans and Peterson, 1987
[c] Fuchsman, 1978
[d] Fuchsman, 1980
[e] Cameron and Schruben, 1983
[f] Mutanen, 1992

Metals content of peat   Emissions of dust and metals are dependent on the amount of ash and metal found in a given fuel. The metal content of peat is given in Table 1.2.

Table 1.2 *Some typical values for the trace elements in fuel peat*

Trace elements	Range µg/g dry peat		Mean µg/g dry peat
Arsenic	0.40	– 17.50	2.90
Beryllium	0.074	– 0.13	0.095
Cadmium	0.40	– 1.40	0.18
Chromium	1.30	– 154.00	5.40
Cobalt	0.26	– 3.10	1.30
Copper	1.40	– 15.90	5.60
Iron	1300.00	– 12800.00	6400.00
Lead	1.10	– 31.30	5.40
Manganese	46.00	– 75.00	60.00
Mercury	0.00	– 1.11	0.35
Molybdenum	0.20	– <4.80	2.00
Nickel	1.20	– 23.50	3.60
Vanadium	0.008	– 3.33	6.20
Zinc	2.60	– 66.60	11.80

(Source: Aittola, 1992b)

Trace gas emissions from peatland   A large deposit of peat will emit nitrous oxide (N_2O), CO_2 and methane at low but fairly constant levels. Climate warming will have a negative effect on peatland emissions through increasing drainage and converting what was once a sink for CO_2 into a source of CO_2 (Moore, 1994).

Coal

There are no universally agreed definitions for different categories of coal. Based on a variety of criteria, coal can be grouped with respect to type, rank and grade.

Coal type   All coals are known as humic fuels since they are derived from plant materials. The determination of coal type requires a microscopic evaluation of the coal in question to pinpoint the maceral group to which it belongs.

The maceral groups are defined according to the characteristics of the plant material from which the coal is formed.The three major maceral groups are exinite (spores, resins, waxes), vitrinite (wood) and fusinite (carbonized wood).

Coal rank The transformation of plant material over time, under conditions of high pressure and high temperature, from oxygen-rich plant debris to carbon-rich, oxygen-poor anthracite coal is known as coalification (see Figures 1.2 and 1.3). The degree of coalification generally, but not always, relates to age, with the youngest coal being lignite and the oldest coal, anthracite. Low-rank coal is distinguished from higher-ranked coals by a high moisture content, low heating value, high oxygen content and a generally alkaline but highly variable inorganic content (Sondreal, 1992).

Two types of analysis, proximate and ultimate, can be carried out to determine coal rank. Proximate analysis involves standard tests to determine the volatile matter, the equilibrium moisture present at 105°C, the ash residue and the fixed carbon of a coal sample. Ultimate analysis provides information on the carbon, hydrogen, sulphur and nitrogen present in dry, ash-free coal.

Coal grade The grade of a coal is identified according to its value for a given end use. Factors which can contribute to grade include calorific value, as well as ash, moisture, oxygen and sulphur content. The end use of a high-grade fuel (as mined) is strongly influenced by its mineral matter content. Therefore, high-rank coal which contains a large quantity of ash may be considered a low-grade fuel even if its ash-free calorific value is high. The suitability of coal as a boiler fuel is also principally determined by its sulphur and ash content as well as the combustion properties of the ash, rather than rank (Sondreal, 1992).

The quantity of ash present in a mined quantity of coal can be influenced by the chosen mining method (e.g. underground mechanical mining often produces a higher ash coal than surface mining). This means that there is no reliable way of grading a coal supply by its rank.

Lignite Lignite is generally the least energy efficient of all the coals with a calorific value, as mined, in the range of 6–15.9 MJ/kg (1434–3824 kcal/kg) (Couch, 1988; Monenco, 1981). For this reason lignite will be the primary focus of this study among the fuels from the coal group. However, there is a percentage of higher-ranked coals that, due to a high impurity content, may also fall within the definition of low-grade fuels used here. Where information regarding these higher calorific value low-grade fuels is available with respect to location, environmental implications,

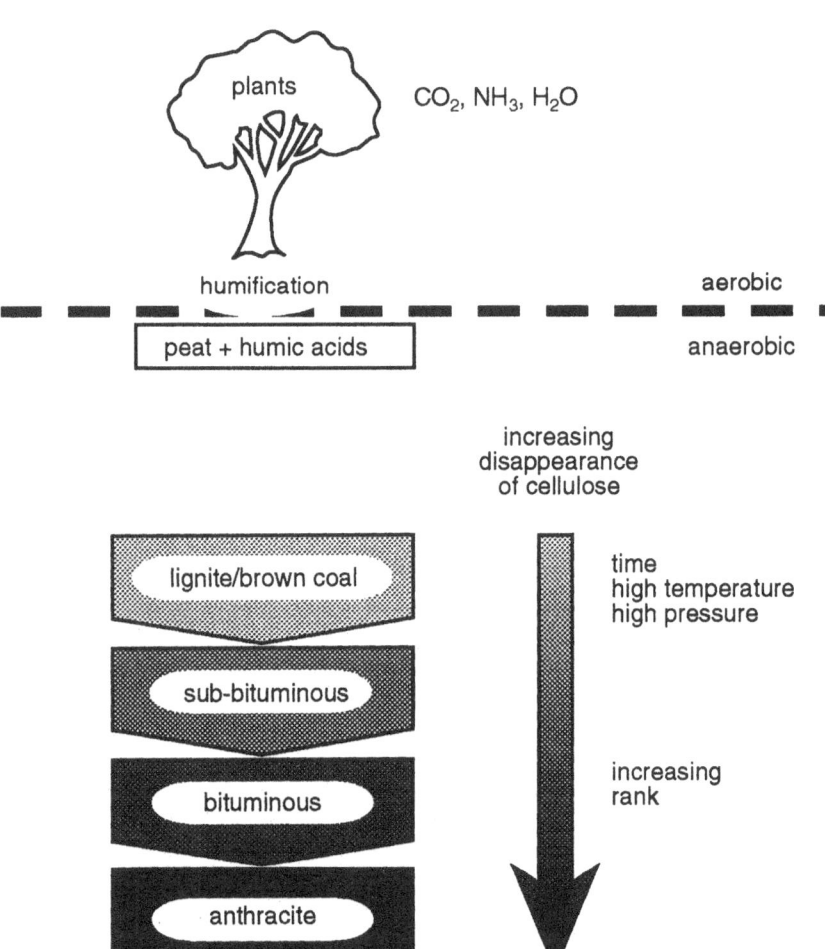

Figure 1.3 *The coalification sequence*

(Source: Adapted from Allardice et al., 1981)

technology for use and mitigation of impacts, it will be included in the discussion. Lignite itself is also referred to as brown coal, particularly in Eastern Europe. For the discussion presented here the term 'lignite' will be used to indicate lignite and brown coal.

Like peat, lignite is a fossil fuel that is formed over a long period of time from the breakdown of organic material, primarily plants. As Figure 1.3 illustrates, it is the least dense and least decomposed of the various ranks of coal. The factors which determine the properties of different types of coal, for example, carbon content, moisture content, volatiles and sulphur content, are not discrete, so there are some sub-bituminous coals which have a lower calorific value than the highest carbon content lignites.

Properties of lignite Lignite is formed under a variety of conditions, which leads to great variability in the properties of lignite both within and among deposits. Couch (1988) has indicated that this variability is 'characteristic' of lignite particularly when the quantity of ash and the properties of that ash are under consideration. The sulphur content of lignite varies widely but is a concern from an environmental standpoint, since a sulphur content in the range of 1–3 per cent by weight is average and can be as high as 11.5 per cent (Mutanen, 1992; Monenco, 1981). As the calorific content of lignite is relatively low, high-sulphur lignites may be the dominate source of sulphur emissions in countries where lignite has been exploited. An overview of the properties of lignite is presented in Table 1.3.

Table 1.3 *Properties of lignite*

Property	Range %
Carbon	65.0–78.0
Hydrogen	4.5– 9.7
Oxygen	11.0–30.0
Nitrogen	0.4–30.0
Sulphur	0.3–11.5
Ash	2.0–49.0
Moisture	15.0–60.0
[Volatiles	45.0–60.0]

(Source: Mutanen, 1992; Ots, 1992; Tigges, 1990; Couch, 1988; Monenco, 1981)

Metals content of lignite As with the other properties of lignite, metal content differs greatly according to locality. Table 1.4 gives a general metal content for all coals including lignite.

Table 1.4 *Some typical values for the trace elements in lignite*

Trace elements	µg/g
Arsenic	1.40– 7.80
Beryllium	0.50–14.60
Cadmium	0.02– 0.30
Chromium	3.20–12.70
Lead	0.80– 1.10
Mercury	0.05– 0.07
Nickel	1.30– 5.50
Selenium	0.24– 1.30
Vanadium	4.20–15.00
Zinc	2.00– 7.50

(Source: Heinrichs,1982; 1977; Schwitzgebel et al., 1975)

Oil shale

Oil shale, prior to processing, may be difficult to combust, but it is a potential fuel since its calorific value is approximately 3.1–8.7 MJ/kg (750–2100 kcal/kg) as mined (Ekinci, 1995; Kikas, 1992; Roumani, 1992). The way in which oil shale formation takes place makes it high in elements which have a number of uses other than for fuel (Aarna, 1978). These alternative uses are discussed in Chapter 2.

Oil shale was formed underwater from small sea plants and animals which thrived in warm temperatures and settled on the sea bottom after their death hundreds of millions of years ago. Oil shale deposits contain alternate layers of oil shale and limestone. The presence of limestone in these deposits is thought to serve as an indication of times of lower temperature in the sea when lime bacteria activity increased and thus limestone formation occurred. Both coal formation (carbonization) and shale formation (bitumenization) take place in oxygen-free (anaerobic) conditions. Coal formation results from high pressure and high temperatures acting on organic material over long periods of time, while oil shale formation is the outcome of the transformation of organic material underwater in relatively lower temperatures (Aarna, 1978).

Properties of oil shale The composition of oil shale, like other low-grade fuels, varies greatly. It differs from other fuels in that it contains many chemical compounds that have applications outside fuel use. Oil shale is not yet widely used as a fuel. For this reason, and due to the fact of its chemical complexity, there are still uncertainties surrounding the properties of oil shale. Table 1.5 presents information regarding some general characteristics of oil shale.

Table 1.5 *Properties of oil shale*

Property	%
Carbon	9.50–19.00
Hydrogen	1.09–1.80
Oxygen	0.13–11.00
Nitrogen	0.06–3.00
Sulphur	0.60–9.30
Ash	5.00–80.00
Moisture	2.00–20.00
[Volatiles	85.00]

(Source: Roumani, 1992; Castaneda Perez, 1992; Hiltunen, 1990; Tigges, 1990; Aarna, 1978)

Metals content of oil shale Some typical quantities of trace elements found in oil shale are presented in Table 1.6.

Wood

The calorific value of wood can range from approximately 6.2–12.7 MJ/kg (1500–3040 kcal/kg) (Mutanen, 1992; Monenco, 1981). The calorific value of wood is related to its moisture content as well as the type of wood. Wood is similar to other low-grade fuels in that there is a great deal of variation with respect to its various properties. Soft woods like *Eucalyptus*, whose cells are more loosely packed, often have a lower energy density than hardwoods such as ebony and mahogany, whose cells are much closer together.

Table 1.6 *Some typical values for the trace elements in oil shale*

Trace elements	µg/g
Arsenic	7.00– 40.00
Beryllium	35.00
Cadmium	0.14
Cobalt	5.00– 39.00
Chromium	34.00– 280.00
Copper	15.00– 112.00
Fluorine	968.00–1700.00
Mercury	<0.10– 0.20
Molybdenum	5.00– 94.00
Nickel	11.00– 138.00
Lead	10.00– 28.00
Strontium	69.00– 790.00
Thorium	0.77
Thallium	<0.14– 0.40
Uranium	0.99– 27.00
Vanadium	29.00– 760.00
Zinc	13.00– 444.00

(Source: Robl,1983; Shirav and Ginzberg, 1983; NAS, 1979)

The calorific value of wood is also related to its moisture content. Green saplings will not burn easily, and older wood requires drying to increase its calorific value. As wood dries naturally after cutting, the properties of wood change over time. Fuelwood is often stored for a time, prior to use, for precisely this reason.

In Europe wood for domestic fuel use may be harvested from plantations or wood lots specifically managed for that purpose, although in many developing countries fuelwood more often consists of dead branches and twigs collected by hand. The moisture content of the many varieties of wood that are gathered for fuel around the world is difficult to predict, primarily because the amount of time that this wood will have spent drying in the air and sun is unknown.

Properties of wood   An overview of the chemical and physical characteristics of wood is given in Table 1.7.

Metals content of wood   Table 1.8 indicates some typical values of the trace elements in wood.

Table 1.*7 Chemical and physical characteristics of wood*

Property	%
Carbon	48.00–51.00
Hydrogen	6.00– 6.50
Oxygen	38.00–42.00
Nitrogen	0.30– 2.30
Sulphur	<0.05
Ash	0.40– 0.60
Moisture	25.00–55.00
[Volatiles	75.00–88.00]

(Source: Mutanen, 1992; Monenco, 1981)

Table 1.8 *Some typical values of the trace elements in wood*

Trace elements	μg/g
Arsenic	0.01– 0.42
Beryllium	–
Cadmium	<0.19– 0.59
Chromium	0.90
Cobalt	0.13
Copper	0.55– 2.40
Iron	8.90– 52.00
Lead	<1.00– 13.00
Manganese	22.00–754.00
Mercury	0.02– 0.18
Molybdenum	<2.10– 5.40
Nickel	0.57
Selenium	0.19
Vanadium	1.90
Zinc	6.10–118.00

(Source: Aittola, 1992b)

Biomass other than wood

This group of low-grade fuels includes a wide variety of substances such as crop residues (shaft, stocks, leaves), nut shells, dung and dried water plants. Biomass other than wood is generally used as a fuel by poorer members of a community in an area where wood is scarce or unavailable. As with all low-grade fuels the characteristics of these fuels are highly variable, and very little work has been carried out regarding their composition.

It has been estimated that up to 30 per cent of Turkey's annual energy consumption could be supplied by agricultural residues, replacing all the lignite and bituminous coal used in generating electricity. This could have a positive effect on the environment, since the SO_2 and NO_x content of crop residues is generally lower than lignite (Ergudenter, 1994).

The calorific values of agro-residues can be as high as 23.3 MJ/kg (Eriksson and Prior, 1990), although other forms of biomass (e.g. dung 6–11 MJ/kg) can be significantly lower. Briquetting of these materials will increase their calorific value by volume. In addition, it may also lead to the buying and selling of these substances. In an 'as collected' state these fuels are often considered a free good.

Municipal solid waste

Household waste and commercial waste are commmonly known as municipal solid waste (MSW), since it is often the responsibility of the local government authority to provide for the disposal of these wastes. The calorific value of municipal waste is extremely variable, from 4–15 MJ/kg (956–3585 kcal/kg) or even higher, depending upon the composition of the waste. These wastes may be burned in an 'as collected' state or they may undergo processing (e.g. sorting, compaction) to achieve a fuel with more uniform characteristics. Processed MSW may be referred to as refuse-derived fuel (RDF). A binder may also be added to produce powder RDF which is generally more biologically and chemically stable than unprocessed MSW (Daugherty, Ohlsson and Venables, 1988).

Incineration has been one of the traditional methods of refuse disposal. In Europe, it is increasingly the case that to compete with the cost of landfill, incineration must be carried out for the purposes of power production and not for disposal alone. The combustion of MSW and RDF can lead to the emission of more volatile trace metals such as mercury, cadmium and lead (Clayton and Scott, 1986; Scott et al., 1986). Some typical values for the trace elements content of MSW are given in Table 1.9.

Even MSW which has been sorted is likely to contain plastics and other chlorinated compounds which, when burned, can produce toxic substances such as dioxins and furans. TCDD (tetrachlorodibenzo-*p*-dioxin) is very toxic to plants and animals and is known to contribute to chromosome

Table 1.9 *Some typical values for the trace elements in MSW*

Trace elements	µg/g
Cadmium	2.50–5.50
Chromium	7.00–10.00
Cobalt	2.00–7.00
Copper	100.00–1500.00
Lead	50.00–1600.00
Manganese	400.00–1400.00
Mercury	0.50–30.00
Nickel	5.00–30.00
Zinc	600.00–1600.00

(Source: Aittola, 1992b, 1987)

malformation (OECD, 1988). TCDD may also be present in waste prior to combustion if products containing the substance (e.g. waste from manufacturing the herbicide 2,4,5-T) are included in the MSW. Table 1.10 presents information on ground level concentrations of organic chlorine and dioxin from two sources in the United States that burn waste. Flue gas cleaning using lime and dust collection using fabric filters can reduce dioxin emissions to as little as one-hundredth of the untreated discharge (Statens Energiverk, 1987).

Global use of low-grade fuels

Low-grade fuels contribute to the energy mix of many countries. A detailed discussion of reserves and resources of low-grade fuels is provided in Chapter 3. A brief overview is given here.

Peat

Peat is currently harvested in approximately 15 countries in both milled (shredded) and sod (bulk) form. The World Energy Conference (1989) reports that in 1987 peat was consumed in 16 countries. Peat is relatively widely used as a fuel in only four countries, the former Soviet Union, Sweden, Ireland and Finland. Its use in power production is growing (Lehtovaara, Nieminen and Kosunen, 1990). In most countries where peat is harvested, it is used as a soil supplement or growing medium for agricultural and horticultural purposes (WEC, 1989; Boron, Evans and Peterson, 1987).

Table 1.10 *Estimated highest mean annual ground level concentrations of organic chlorine (TOCl) and dioxin emissions*

	Co-firing[a]	Mass incineration[b]
TOCl, flue gas, % input	0.6%	3.0%
TOCl, flue, ng/m³	616.00	3200.00
Dioxins, ng/m³	No data	505.00
2,3,7,8-TCDD, ng/m³	No data	0.41
Tetra-CDD*, ng/m³	No data	6.30
Tetra-CDF**, ng/m³	90.00	No data

$$TCDD/TCDF = 96 \, ng/m^3 \times 1.3 \times 10^{-6} \, (atmospheric \, dilution \, factor)$$
$$= 1.25 \times 10^{-6} \, g/m^3$$

* Tetrachlorinated dibenzodioxin
** Tetrachlorinated dibenzofuran
[a] Ames, Iowa , coal/RDF co-firing unit burning 10% RDF, 90% coal
[b] Chicago, Illinois, waterwall incinerator burning 100% MSWs

(Source: OECD, 1988)

Lignite

European consumption of lignite, including countries which have had a centrally planned economy since the early or mid-part of this century, accounts for over 80 per cent of the world total for lignite consumption (WEC, 1989). The countries situated in Eastern Europe have a particularly abundant supply of lignite (Josephson, 1993). During 1987 no lignite consumption was recorded in either Africa or South or Central America although proven recoverable reserves do exist in these areas. In Asia, lignite was used in China, Mongolia, India and Thailand. Lignite consumption was also recorded in Canada, the United States and Iran (WEC, 1989).

The use of lignite as a fuel has increased greatly in the past 20 years in countries in Eastern and Southern Europe, Asia and Australasia. The primary impetus for this is the expansion of its use in the power sector. An increase in the overall use of lignite in Europe is demonstrated by the values in Table 1.11.

Nearly all of the lignite that is consumed is used for energy production. In Asia, approximately 25 per cent of lignite consumption takes place in the home for lighting and cooking (Jianxiong, 1989; Couch, 1988).

Table 1.11 *Lignite production in Europe: 1983–87*

	1983 megatonnes	1984 megatonnes	1985 megatonnes	1986 megatonnes	1987 megatonnes
Austria	3.0	2.9	3.1	3.0	2.8
Bulgaria	32.1	32.1	30.7	34.9	36.0
Czechoslovakia	102.4	104.7	102.3	100.8	100.4
FRG	124.3	126.7	120.7	114.4	108.9
GDR	278.0	296.3	312.2	312.2	319.0
Greece	30.6	32.6	35.9	36.5	42.8
Hungary	22.4	21.5	21.4	20.9	19.7
Poland	42.5	50.4	57.8	64.9	73.2
Romania	30.4	35.8	37.9	38.0	38.0
Spain	24.3	24.4	23.7	22.3	22.4
Turkey	20.7	22.9	25.7	31.5	47.9
Yugoslvaia	58.2	65.1	69.1	72.0	70.0
Others	6.0	5.6	5.8	5.7	5.7
Total	774.9	821.0	846.3	857.1	886.8

Oil shale

Oil shale is not widely used primarily due to the costs involved. Known reserves of oil shale are at least 10 billion tonnes, but at present it is used only in Estonia, China, Brazil and the United States (WEC, 1989). Even in these countries its use is limited to a small number of power plants.

Oil shale can be processed to produce fuel oil which is suitable for use in the transportation and industrial sectors of a nation's economy. It can also provide a wide range of chemical feedstocks for the manufacture of products such as adhesives, resins and plastics (Aarna, 1978).

Wood

Wood is an important source of fuel in many countries, particularly in Africa, Asia and South and Central America. In some economically impoverished countries such as the Sudan or Rwanda, wood often meets up to 90 per cent of domestic energy needs (WEC, 1989; Dankelman and Davidson, 1988). When wood undergoes combustion under oxygen-limited conditions, charcoal is produced. Charcoal is used as a fuel in many countries of the developing world, particularly in the urban centres.

Wood is also used in woodstoves and fireplaces in developed nations, although this accounts for only 1–2 per cent of domestic energy consumption in most of these countries (Pasztor and Kristoferson, 1990; WEC, 1989). Industries which produce wood products, such as pulp and paper manufacturers, can often generate 30–50 per cent of their energy needs by burning waste produced by the plant such as bark and wood residues and combustible liquors (WEC, 1989).

Biomass other than wood

In areas where imported fuels are unavailable and indigenous fuels are scarce, people may turn to biomass fuels to cook food and to provide light (WCED, 1987; IDRC, 1986; Barnard, 1985). Biomass fuels can include woody stems from food crops, woody shrubs, weeds, husks and dung. The practice of using biomass other than wood is especially common in arid regions, such as the countries of the Sahel, but is widely known throughout Asia, Africa, Central America and South America (IDRC, 1986).

Differences in low-grade fuel use between developed and developing countries

There are differences between developed and developing countries with respect to the use of conventional fuels such as oil, natural gas and hard coal. These differences generally relate to the economic constraints experienced by developing countries wishing to purchase expensive high-quality fuels and modern technology for their exploitation. In essence, many of the same forces which determine the differences in energy mix between developed and developing nations influence the current use of low-grade fuels. If the fuel requires expensive shipping and complex recovery and processing equipment, then poorer countries will be unable to use low-grade fuels such as lignite, oil shale or peat as major sources of energy.

The scale of operation is also important to developing countries or regions where money for expensive energy projects and environmental protection is scarce. In remote areas, small- or medium-scale projects may be all that is required. If funding is provided by an international agency for low-grade fuel development, adequate financing should be made available to ensure that proven environmental protection measures are implemented and that the size of the project is appropriate for the desired end use and the context into which the development will be placed.

The transport of low-grade fuel either for processing or consumption is generally more difficult in lesser developed countries than in industrialized nations where rail and road networks are well established. The issue of transport, while it may act as a constraint on the development of low-grade fuels in developing countries, is typical of the differences between developed and developing nations that influence all aspects of development in general.

In the developing nations and in the poorer regions of developed nations (e.g. Newfoundland in Canada or rural Ireland), low-grade fuels such as wood and peat are widely used to meet domestic energy needs. If the low-grade fuel in question can be harvested or collected in a simple manner by an individual and if it is relatively inexpensive, there is a good chance that it will prove to be an attractive fuel for household use in developed and developing regions and countries.

Wood

The only low-grade fuel which is widely used throughout the world is wood. Most of this wood is used to provide energy for light, heating and cooking in many developing countries of the world. Industrial applications for wood are generally limited to countries where the manufacture of wood products takes place and where the wood residue from the manufacturing process can be recycled and burned to provide energy for the factory or for district heating (e.g. Scandinavia).

Lignite

The use of lignite as a fuel is greatest in industrialized countries. Although reserves are well documented in a number of developing countries such as India, Indonesia and China, these have not been exploited to any great extent. One reason for this may lie in the great variability among and within lignite deposits, which necessitates specially designed processing or combustion equipment that reflects the individual character of a specific lignite deposit. The cost of this technology can be prohibitive. There is, however, a growing interest in using lignite in the power sector in countries such as Thailand which have a rapidly growing demand for electricity.

Where high quality fossil fuel alternatives are available in developing countries (e.g. oil in Peru, anthracite and bituminous coal in China) and where it is not essential to export these fuels to ease balance of payment problems, these will obviously be preferred for energy production over lignite.

Peat

A number of industrial countries continue to refine technology for the industrial application of peat as a fuel (e.g. Finland and the former Soviet Union). Most of the peat used in Scotland is hand-cut sod which is used to produce energy in the home. The financial costs associated with the use of peat to produce energy for the industrial sector are too high for some countries. In Burundi, for example, hydropeat technology could be used to exploit 42 million tonnes of peat located in one region. The recovery and use of this peat have been judged to be uneconomical, however, since the market for peat as a fuel is small (WEC, 1989).

As with lignite, large-scale production for the power sector is required to make utilization financially viable. Such large-scale production requires equally large markets. In Finland, for example, peat accounts for approximately 5 per cent of the energy produced.

Oil shale

The economic viability of oil shale is dependent upon the price of oil. Oil shale can be processed to produce oil but the technology required to carry this out is complicated and highly priced. The use of oil shale, therefore, is limited in developed countries due to its high financial cost. For this same reason it is used even less in developing countries.

Current applications of low-grade fuels

There are a number of ways in which low-grade fuels are used to produce energy, including combustion, gasification and liquefaction, which will be outlined briefly here. A more thorough technical discussion and a consideration of the environmental impacts of using low-grade fuels are contained in Chapters 4, 5, 6 and 7.

Combustion

Conventional boilers The application of heat to a material in the presence of oxygen will cause combustion. There is a variety of approaches to the combustion, or burning, of low-grade fuels. For industrial purposes oil shale, lignite, peat and wood can be burned in boilers to produce steam heat for the heating of buildings. More often, the superheated steam is used to provide energy for industrial processes or large district power plants. A traditional boiler design includes a fixed floor at the bottom of the boiler on which the fuel rests (Figure 1.4).

Due to the high degree of variability regarding properties such as moisture, sulphur and ash content in solid low-grade fuels, conventional boilers are not usually the best method for achieving efficient combustion of these fuels. The variability in lignite, for example, can lead to a build-up of ash and residue on the boiler heat transfer surface which eventually interferes with efficient combustion (Goblirsch and Talty, 1980). In addition, conventional boilers do little to control emissions of SO_2 and NO_x.

Experience with burning oil shales in Estonia has indicated that the high mineral component of oil shale, coupled with the properties of the ash which forms during combustion, makes the use of conventional boilers impractical (Ots, 1992). Problems with conventional boilers such as ash smelting and particulate emissions have also been reported with wood and peat in Finland (Mutanen, 1992).

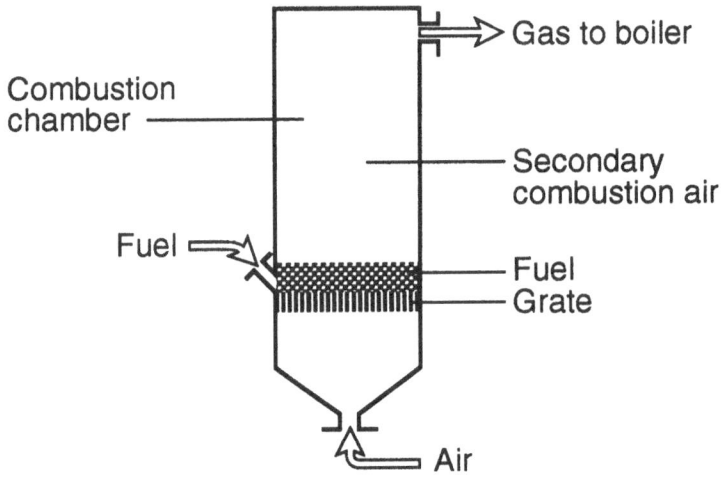

Figure 1.4 *A simplified fixed bed boiler*

Alternatives to conventional boilers   Modifications carried out on conventional boiler design have led to increased efficiency in the combustion of low-grade fuels. The majority of these changes involve the floor of the boiler on which the fuel rests. Fluidized bed boilers, where the fuel is burnt on a bed of turbulent air, have proven particularly successful for the burning of lignite, peat and wood and are able to handle a range of changes in fuel characteristics along with protecting the environment from sulphur and, to a lesser extent, nitrogen oxides (ECE/UNDP, 1986; IDRC, 1986).

There are a number of variations based on the fluidized bed combustion principle, including increasing circulation in the boiler to ensure a greater combustion efficiency due to increased burning of fine particles. Figure 1.5 illustrates a fluidized bed boiler and a circulating fluidized bed boiler.

Pulverizing the fuel prior to combustion has been effective in ensuring dependability of heat production when burning oil shale in the former Soviet Union (Ots, 1992). Installing a moving grate, rather than a fixed bed, has achieved positive results with peat and wood residues in Finland leading to cleaner and more thorough burning of these fuels (Mutanen, 1992). Combinations and variations on alternative combustion techniques discussed here, such as grate-pulverized combustion and pressurized fluidized bed combustion, are currently in use in many countries. These technologies are discussed in more detail in Chapter 7.

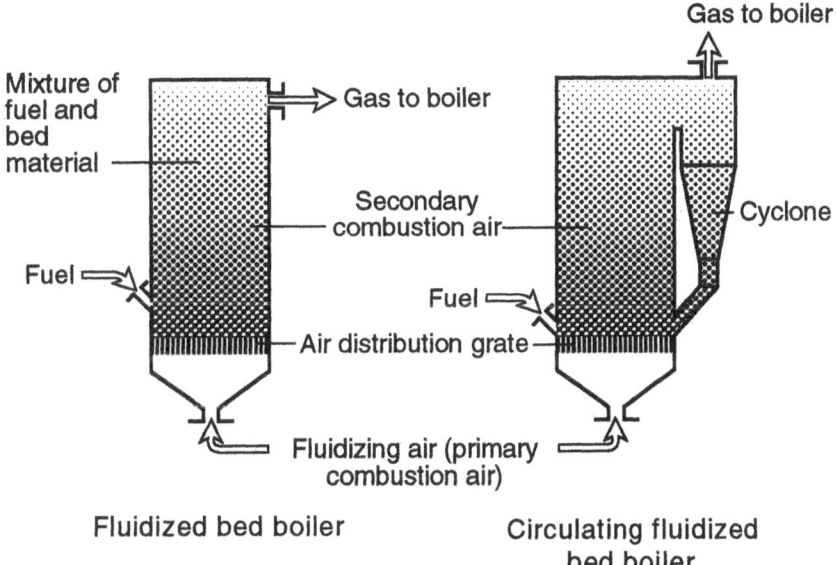

Figure 1.5 *Examples of modifications to conventional boiler design*

Stoves The domestic combustion of low-grade fuels usually takes place in a stove or fireplace. In warmer latitudes the stove will generally be used for cooking while in northern regions and high altitude areas in warmer countries wood and other biomass, and to a lesser extent peat and lignite as received or in pellet, briquette or crude form, will be burned to provide heat as well. There is a great deal of interest in the efficiency of cookstoves in developing countries. This aspect of energy conservation is considered in Chapter 8.

Gasification

The gasification of solid low-grade fuels increases the potential applications of these fuels. Not only can gas from low-grade fuels be used for combustion during power generation, but it can also be used in the production of chemicals and to provide energy for specially equipped transport vehicles (Chadwick, Highton and Lindman, 1987; Kjellström, 1985; Mujadin and Weinrich, 1985; Martins et al., 1980).

Gas produced during carbonization or retorting of a solid fuel is called producer gas, while gas which results from a process designed to gasify a fuel is known as synthetic natural gas (SNG). The product of the gasification of wood and other biomass is generally referred to as biogas. In some instances, these terms are used interchangeably in the literature.

Gasification of solid low-grade fuels can occur in a number of ways. Peat can be converted to a gaseous state by a thermal process such as a fluidized bed reactor (Lau, 1981) or through a biological process such as anaerobic digestion where, under oxygen-free conditions, organic matter breaks down into CO_2 and methane. Thermal gasification of peat requires dewatering to achieve less than 50 per cent moisture, while high moisture content peat can be used in biogasification (Chynoweth, 1983).

The gasification of lignite can occur using a fixed bed process under low pressure where air and steam provide the gasifying vehicle. Fluidized bed and moving bed gasification take place under comparatively higher pressure and are only cost-effective when implemented on a large scale (Harker and Backhurst, 1981). In Germany, hydrogen is added to lignite in a fluidized bed at temperatures of 850 to 900°C. Up to 7800 m^3/h of methane has been produced using this method (ECE/UNDP, 1986).

Combustible gases may be recovered from oil shale using a fluidized bed gasifier (Martins, 1992). Gas is also produced during the semicoking of oil shale. This gas is often recycled back to heat the coking chambers (Valgus, 1974).

Anaerobic digestion of wood and other biomass will produce methane. This process can be carried out in a simple biogas digester to provide fuel for a single household or on a larger scale. Biogasification of low-grade biomass, such as dung or crop residues, will result in a low calorific value fuel.

Liquefaction

The conversion of solid low-grade fuel to liquid fuel enables low-grade fuels to act as liquid substitutes for oil products such as petroleum. In general, liquid fuels are less expensive to handle and transport than solid or gaseous fuels (IDRC, 1986).

Direct liquefaction of peat is still at the experimental stage (Boron, Evans and Peterson, 1987). Indirect liquefaction, where peat is converted to synthetic gas which is then added to a liquid, is a more thoroughly tested method. The product of the indirect liquefaction of peat is methanol, an alcohol, which can also be produced from wood using the technique of prior gasification.

Another alcohol fuel, ethanol, can also be produced from wood and other biomass following fermentation. This process is not as economical as methanol for wide-scale production (IDRC, 1986).

Liquefaction of coal can be carried out directly, for example, through hydrogenation of coal using high temperature and high pressure plus hydrogen, or indirectly by gasifying the coal then synthesizing the coal into a liquid (Chadwick, Highton and Lindman, 1987; ECE/UNDP, 1986; Harker and Backhurst, 1981). Direct liquefaction is a more efficient method of converting coal into a liquid fuel, but it is not easily done on a large scale (Harker and Backhurst, 1981).

The distillation of oil shale will produce crude shale oil.

Environmental impacts of low-grade fuel use

Negative impacts to the environment from the use of low-grade fuels can result from a number of factors which are outlined briefly below. A more detailed assessment of the potential impacts associated with the expanded use of solid low-grade fuels is presented in Chapters 4, 5, 6 and 7.

The fuel cycle

'Fuel cycle' is the term used to describe the steps involved in obtaining and using a fuel. The major stages in this cycle (illustrated in Figure 1.6) are mining (or gaining), processing, transport, conversion and utilization. Other activities may also occur in the fuel cycle, such as storage, and an action such as transport may be required at more than one place in the cycle. Environmental impacts occur at points throughout the cycle. Preventing, eliminating or mitigating these impacts is essential if the negative effect of fuel use on the natural and socio-economic environment is to be minimal. Figure 1.7 provides an overview of the potential environmental effects throughout the fuel cycle associated with using coal to produce energy.

Scale of operation

The scale of operation is an important consideration in determining the potential impacts to the environment of a fuel cycle. The size of the mine or gaining area, the desired end use, which can dictate boiler size (e.g. heating, steam production or power generation), economic viability and demand are several indicators of the environmental impact/scale relationship. For example, the boiler size can influence the scale of mining which will be required and, therefore, the degree of environmental disruption or degradation which will result.

Environmental standards

Environmental standards and regulations are not uniform from country to country. Existing laws and public attitudes towards environmental

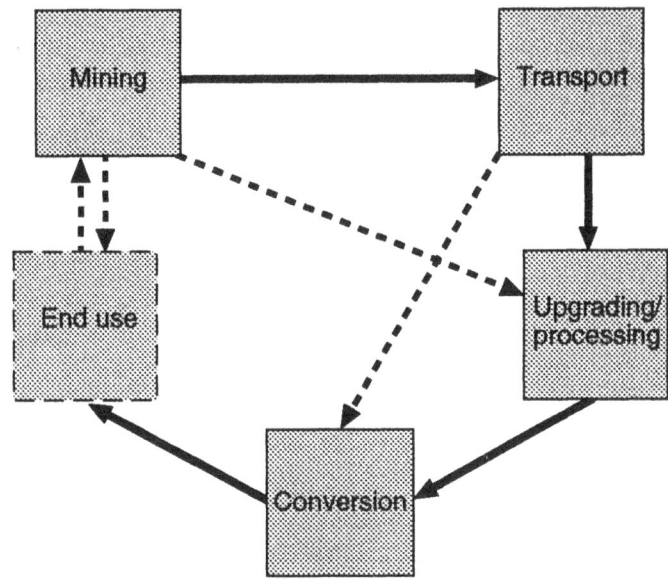

Figure 1.6 *Relationships of the major stages in the fuel cycle*

protection can also contribute to the environmental impacts of low-grade fuel use through the enforcement of strict control measures or the allowance of environmental damage.

The aim here is to provide information on technologies and policy options which will ensure environmentally responsible use of low-grade fuels. As reports such as *Our Common Future* (WCED, 1987) illustrate, there is worldwide awareness of and concern for the state of the environment which must be taken into account in any expanded use of low-grade fuels in the present and the future.

Conservation

The adoption of a conservation plan can reduce the amount of fuel required for a specific end use. Fuel efficient boilers, for example, can reduce the energy demand for a given fuel over the long run, which can have a positive effect on the environment throughout all stages of the fuel cycle. This concept is explored more fully in Chapter 8.

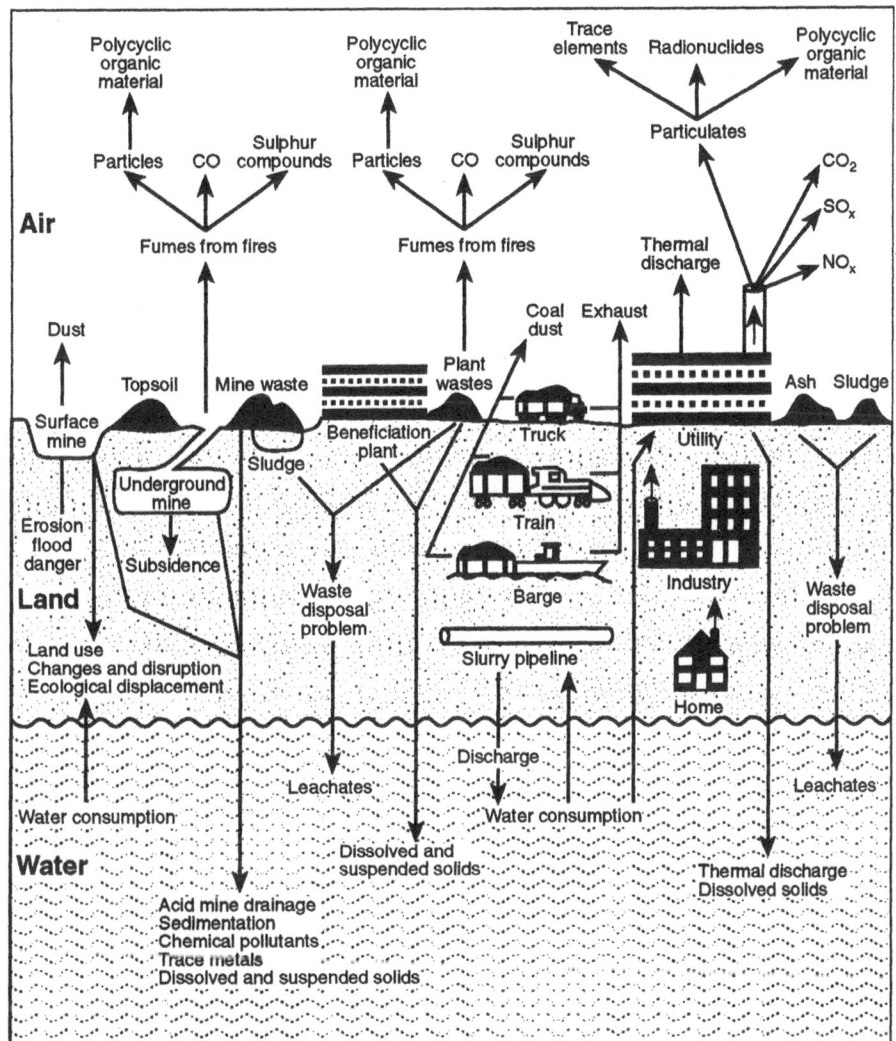

Figure 1.7 *Environmental disturbances from coal-related activities*

(Source: US Congress, Office of Technology Assessment, 1979)

Technology transfer

Technologies currently exist for controlling or preventing many of the negative environmental impacts associated with the combustion of low-grade fuels. The vast majority of these technologies, including the actual machinery as well as information such as patents, designs, copyrights and trade secrets, which are collectively known as intellectual property, belong to large corporations of the North. In general, the market principles of these companies are at odds with simply giving the environmental protection technology to developing countries or European economies in transition.

Appreciation that many environmental problems extend beyond national borders, the most recent being the problem of global warming, has created an atmosphere of international co-operation which has extended to the search for methods of achieving the transfer of environmentally sound technology. Some of the approaches to technology transfer which are currently in use or under consideration and which are relevant to the environmental issues surrounding low-grade fuels are examined in Chapter 9.

Environmental impacts associated with individual low-grade fuels

Peat Harvesting of peat will destroy the vegetational cover of a peatland. This cover may contain rare or important plant species. Following peat harvest, it may be possible to revegetate a peatland, although creating the original ecological diversity may be difficult. The ground and surface water of an area may also be disrupted during and following the harvesting of peat due to the high moisture content of most peats. The degree of impact caused by peat harvesting is related to scale. Harvesting in a small area will usually cause fewer impacts than large-scale harvesting (Boron, Evans and Peterson, 1987).

The combustion of peat generally produces emissions that are low in sulphur and NO_x although this can be dependent on the technology employed. Carbon dioxide is also emitted during peat combustion.

Lignite Negative impacts to the environment may occur during the mining, transport and combustion of lignite. Opencast mining is the most commonly used mining technique to gain lignite both because it is a cheaper alternative to underground mining and because lignite deposits tend to be closer to the surface than deposits of older coals.

During open pit mining the ground cover of the area mined will be destroyed and dust and noise may occur due to the use of heavy equipment and blasting. Impacts to individuals living in the vicinity of an open pit mine and along transportation routes may include dust, noise and vibration, which can have an impact on activities of daily living for area residents. The degree to which this occurs depends on factors such as age, health and length of time at home during the day as well as the effectiveness of measures implemented to mitigate these effects.

The combustion of lignite without control mechanisms may release relatively large amounts of sulphur, nitrogen and CO_2 into the atmosphere along with volatiles, dust and trace metals which may be present in the fuel.

Oil shale Oil shale is mined using strip-mining techniques which, similar to lignite, require the removal of ground cover and can cause disruptions to residents in the area due to increased levels of dust, noise and vibration. The build-up of solid waste from oil shale mining and processing, in part due to the presence of limestone in alternating layers with the oil shale, can be considerable. Processing of oil shale can produce large quantities of contaminated water (e.g. waters containing tar and phenols) as well as dust and solid waste.

The combustion of oil shale can lead to emissions of sulphur and nitrogen and, depending upon the temperature of combustion and the oil shale contents, polycyclic aromatic hydrocarbons (PAHs) may be produced. Oil shale combustion emits the largest quantities of CO_2 of all the low-grade fuels under consideration.

Wood The environmental impacts from the use of wood as a fuel are primarily dependent upon the intensity and frequency with which the wood is harvested. In regions where wooded land is scarce, intense collection of wood for fuel can quickly deplete wildlife habitat, contribute to erosion and, ultimately, threaten future availability of fuelwood. Wood taken from plantations which are maintained in order to supply fuel has little potential negative impact on the physical environment.

Mechanized harvesting and transport of wood may cause disruption to residents in the vicinity of a collection site due to dust and noise. The emission of atmospheric pollutants such as SO_2 and NO_x is minimal during the combustion of wood, although CO_2 is produced.

Biomass other than wood The use of crop residues and dung as fuel has few direct negative impacts on the environment. These fuels have a low sulphur and ash content and can generally be collected and used with little disruption to nearby communities. Indirectly, the use of these materials as fuel can potentially impact the local economy and the level of food production if they are used as fuels instead of being applied as soil enhancers or fertilizers.

Municipal and industrial wastes   Municipal and industrial wastes generally refer to liquid and solid wastes from industries (e.g. clarifier sludge from pulp and paper mills) or municipalities (e.g. combustible solid domestic waste) which can be burned to provide energy. The use of these wastes as a fuel must be carefully controlled, because they may contain numerous chemicals and heavy metals which are by-products of product manufacture. In the case of domestic waste, harmful compounds can form during combustion of certain goods such as plastics. Some harmful substances (e.g. dioxins) are actually destroyed during combustion at very high temperatures (greater than 800°C) or reformed in a 'de novo synthesis' in the boiler at 200–450°C. The components of any given industrial or municipal waste must be thoroughly investigated before suitability as a fuel can be determined.

Certain industrial wastes (e.g. untreated bark or sawdust) can be disposed of while providing energy for the industry from which they originated.

References

AARNA, A (1978) *Chemical Engineering in the Estonian SSR.* Perioodika, Tallinn.

AITTOLA, J P (1992a) 'Emissions from low-grade fuel utilization'. In *The Environmentally Sound Management of Low-grade Fuels.* SEI, Stockholm.

AITTOLA, J P (1992b) *Personal Communication.* Manager, Environmental Technology, Viatek, Espoo.

AITTOLA, J P (1987) 'The theory and techniques of particle measuring methods'. *PPP-Seminar, Helsinki.* University of Technology, Espoo.

ALLARDICE, D J, GEORGE, A M, PERRY, C J and ELVIS, C (1981) *The Chemistry of Brown Coal.* State Electricity Commission of Victoria, Melbourne.

BARNARD, G W (1985) 'The use of agricultural residue as a fuel'. *Ambio* **4–5**: 259–266.

BCRA (1987) 'Beneficiation of low-rank coals'. *BCRA Quarterly* **15**: 41–66.

BELOSELSKY, B S, SOLYAKOV, V K and POKROVSKY, V N (1992) 'Environmental problems associated with the utilization of low-grade power-generating fuels'. In *The Environmentally Sound Management of Low-grade Fuels.* SEI, Stockholm.

BORON, D J, EVANS, E W and PETERSON, J M (1987) 'An overview of peat research, utilization and environmental considerations'. *International Journal of Coal Geology* **8**: 1–31.

CAMERON, C L and SCHRUBEN, P (1983) 'Variations in mineral matter content of a peat deposit in Maine resting on glacio marine sediments'. In *Proceedings of a Workshop on Mineral Matter in Peat.* Los Alamos.

CASTANEDA PEREZ, M (1992) 'Oil shale deposits in Mexico'. In *The Environmentally Sound Management of Low-grade Fuels*. SEI, Stockholm.

CHADWICK, M J (1992) 'Environmental implications of the use of low-grade fuels'. In *The Environmentally Sound Management of Low-grade Fuels*. SEI, Stockholm.

CHADWICK, M J, HIGHTON, N H and LINDMAN, N (1987) *Environmental Impacts of Coal Mining and Utilization*. Pergamon Press, Oxford.

CHYNOWETH, D (1983) 'A novel process for biogasification of peat'. *Proceedings of the International Symposium on Peat Utilization*. Bernidjl.

CLAYTON, P and SCOTT, D W (1986) *The Measurement of Suspended Particles, Heavy Metal and Selected Organic Emissions at Sheffield Municipal Refuse Incinerator*. Warren Spring Laboratory, Stevenage.

COUCH, G R (1988) *Lignite Resources and Characteristics*. IEA Coal Research, London.

DANKELMAN, I and DAVIDSON, J (1988) *Women, Environment and the Third World: Alliance for the Future*. Earthscan, London.

DAUGHERTY, K E, OHLSSON, O O and VENABLES, B J (1988) 'Emission studies of full-scale co-firing of pelletized RDF/coal'. *Conference on Energy from Biomass and Waste XII*. New Orleans.

ECE/UNDP (1986) *Low-Calorie-Value Solid Fuels: Combustion, Gasification, Liquefaction*. United Nations, Geneva.

EKINCI, E (1995) *Economic Considerations of Oil Shale and Related Conversion Processes in Composition, Geochemistry and Conversion of Oil Shales*. Kluwer Academic Publishers, Dordrecht.

ERIKSSON, S and PRIOR, M (1990) *The Briquetting of Agricultural Wastes for Fuel*. FAO, Rome.

ERGUDENTER, A (1994) 'Agricultural residues as a potential resource for environmentally sustainable electric power generation in Turkey'. *Renewable Energy* **5**: 786–790.

FUCHSMAN, C H (1980) *Peat: Industrial Chemistry and Technology*. Academic Press, New York.

FUCHSMAN, C H (1978) *The Industrial Chemical Technology of Peat*. Minnesota Department of Natural Resources, St. Paul.

GOBLIRSCH, G M and TALTY, R D (1980) 'Recent developments in fluidized bed combustion technology for lignites and sub-bituminous coals in the USA'. *Fluidized Combustion: Systems and Applications*. Series 4. Institute of Energy Symposium, London.

HAGELÜGEN, M (1988) 'Brown coal mining in Europe in 1987'. *Glückauf* **124**: 379–390. English translation.

HAMILTON, L D (1992) 'Health and environmental management of low-grade fossil fuels'. In *The Environmentally Sound Management of Low-grade Fuels*. SEI, Stockholm.

HARKER, J H and BACKHURST, J R (1981) *Fuel and Energy*. Academic Press, London.

HEINRICHS, H (1982) 'Trace element discharge from a brown coal-fired power plant'. *Environmental Technology Letters* **3**: 127–136.

HEINRICHS, H (1977) 'Emissions of 22 elements from brown coal combustion'. *Naturwissenschaften* **64**: 479–481.

HILTUNEN, M (1990) 'SO$_2$, NO$_x$ and halogens emission control in pyroflow circulating fluidized bed boilers when using low-grade fuels'. In *VTT Symposium 108: Low-grade Fuels*. Technical Research Centre of Finland, Espoo.

HYÖTY, P (1992) 'Firing of fuels with low calorific value and fluidized bed boilers'. In *The Environmentally Sound Management of Low-grade Fuels*. SEI, Stockholm.

IDRC (1986) *Energy Research: Directions and Issues for Developing Countries*. International Development Research Centre, Ottawa.

IEA (1983) *Concise Guide to World Coalfields*. IEA Coal Research, London.

JIANXIONG, M (1989) *Environmental Problems Caused from Coal Combustion in China and the Countermeasures for Reducing the Effects*. Isinghua University, Beijing (unpublished).

JOSEPHSON, J (1993) 'Eastern Europe's energy and environment'. *Environmental Science and Technology* **27**: 1746–1750.

KIKAS, V H (1992) 'The use of kukersite oil shale ash'. In *The Environmentally Sound Management of Low-grade Fuels*. SEI, Stockholm.

KJELLSTRÖM, B (1985) 'Biomass gasifiers for energy supply to agriculture and small industry'. *Ambio* **14**: 267–274.

LAPPALAINEN, E (1992) 'Geological research on peat resources'. In *The Environmentally Sound Management of Low-grade Fuels*. SEI, Stockholm.

LAU, F S (1981) 'Single-stage gasification'. *Symposium on Peat as an Energy Alternative*. Arlington.

LEHTOVAARA, J, NIEMINEN, M and KOSUNEN, P (1990) 'The quality of fuel peat'. In *Low-grade Fuels*. Vol 2. Technical Research Centre, Finland.

LIEE (1994) *Profiting from Low-grade Heat: Thermodynamic Cycles for Low-temperature Heat Sources*. London Institute of Electrical Engineers, London.

MARTINS, A A (1992) 'Two-stage burning of oil shale'. In *The Environmentally Sound Management of Low-grade Fuels*. SEI, Stockholm.

MARTINS, A A, KARTAU, Y K, NURK, A A and UUERSOO, R N (1980) *Fluidized Bed Oil Shale Gasification Unit* (in Russian). USSR Patent No. 709938. Bulletin 2.

MINNESOTA DEPARTMENT OF NATURAL RESOURCES (1984) 'Inventory of peat resources: an area of Beltrami and Lake of the Woods Counties, Minnesota'. *The Peat Inventory Project*. Hibbing, Minnesota.

MONENCO (ONTARIO LIMITED) (1981) 'Evaluation of the potential of peat in Ontario: energy and non-energy uses'. Occasional Paper No. 7.

MOORE, T R (1994) 'Trace gas emissions from Canadian peatlands and the effect of climate change'. *Wetlands* **14**: 223–228.

MUJADIN, M J and WEINRICH, G N (1985) *The Great Plains Success Story*. ANG Coal Gasification Company, Beulah (Reprint).

MUTANEN, K (1992) 'Energy production from peat in Finland and in the tropics'. In *The Environmentally Sound Management of Low-grade Fuels*. SEI, Stockholm.

NAS (1979) *Redistribution of Accessory Elements in Mining and Mineral Processing: Part II*. National Academy of Sciences, Washington.

OECD (1988) *Environmental Impacts of Renewable Energy: The OECD Compass Project*. Organization for Economic Co-operation and Development, Paris.

OTS, A A (1992) 'The use of Estonian oil shales for power generation'. In *The Environmentally Sound Management of Low-grade Fuels*. SEI, Stockholm.

PASZTOR, J and KRISTOFERSON, L (1990) *Bioenergy and the Environment*. Westview Press, Boulder.

ROBL, T L (1983) 'Geochemistry of oil shales in eastern Kentucky'. In *Geochemistry and Chemistry of Oil Shales* (ed by F P Miknis and J F McKay). American Chemical Society Symposium Series 230, Washington.

ROUMANI, A (1992) 'Prospects for the utilization of oil shales in Syria'. In *The Environmentally Sound Management of Low-grade Fuels*. SEI, Stockholm.

SCHWITZGEBEL, K, MESEROLE, F B, OLDHAM, R G, MAGEE, R A, MESICH, F G and THOEM, T L (1975) 'Trace elements discharge from coal-fired power plants'. *International Conference on Heavy Metals in the Environment: Proceedings*. Toronto.

SCOTT, D W, WOODFIELD, M J, BUSBY, B and WEBB, K (1986) *The Measurement of Suspended Particle, Heavy Metal and Selected Organic Emissions at the Coventry Municipal Refuse Incinerator*. Warren Spring Laboratory, Stevenage.

SHIRAV, M and GINZBERG, D (1983) 'Geochemistry of Israeli oil shales'. In *Geochemistry and Chemistry of Oil Shales* (ed by F P Miknis and J F McKay). American Chemical Society Symposium Series 230, Washington.

SONDREAL, E A (1992) 'Clean utilization of low-rank coals for low-cost power generation'. *Conference on Clean and Efficient Use of Coal: The New Era for Low-rank Coal*. IEA/OECD, Paris.

STATENS ENERGIVERK (1987) *Energy from Waste*. Stockholm.

TIGGES, K D (1990) 'Status of fluidized bed combustion: experiences with circofluid plants in operation'. *Low-grade Fuels*. Vol 1. Technical Research Centre of Finland, Espoo.

US CONGRESS, OFFICE OF TECHNOLOGY ASSESSMENT (1979) *The Direct Use of Coal: Prospects and Problems of Production and Combustion* (OTA-E-86). National Technical Information Service, Springfield.

VALGUS (1974) *The V I Lenin Oil Shale Processing Complex*. Valgus Publishing House, Tallinn.

WCED (1987) *Our Common Future*. Oxford University Press, Oxford.

WEC (1989) *1989 Survey of Energy Sources*. World Energy Conference, London.

2

Alternative uses of low-grade fuels

If low-grade fuels are available, the possibility exists that applications other than energy production will be found for these fuels. As with low-grade fuel use, the environmental implications for alternatives such as chemical feedstock derivation or soil enhancement must be clearly assessed and any negative effects mitigated or eliminated.

The information presented in this chapter is in summary form. For those wishing to initiate a more in-depth study in this area see, Payne, 1987; Hamerslag, 1985; Payne, 1985; Stratton, 1983; Falbe, 1982; Elliot, 1981; Shreve, 1967; Perry, 1963; Lowry, 1945; and Klar, 1925.

Chemical feedstocks

Chemical feedstocks can be derived from peat, lignite, oil shale, wood and municipal and industrial waste. The composition of each of these low-grade fuels varies from location to location reflecting the variety of materials under discussion and, with respect to peat, lignite and oil shale, the conditions under which each deposit was formed. For this reason the use of these substances to provide chemical feedstocks often requires an evaluation of several samples from an individual store or deposit to determine the type of chemicals available for extraction or production.

Knowledge concerning the derivation of chemical feedstocks from low-grade fuels, and their wide-ranging applications, is not complete. Oil shale, in particular, has a complex chemistry which is not yet well understood.

The isolation, or fractionation, of a specific group of chemical feedstocks from substances such as peat, coal, oil shale or municipal waste generally requires a number of steps involving varying degrees of complexity or specialized information. An example of one such process, the isolation of humic acids from peat, can be seen in Figure 2.1.

Chemical feedstocks may be obtained through a number of methods including the use of solvents (to break down chemical bonds), adsorbents (to hold or collect the desired substances), hydrolysis (decomposition of a substance by the chemical action of water) and pyrolysis.

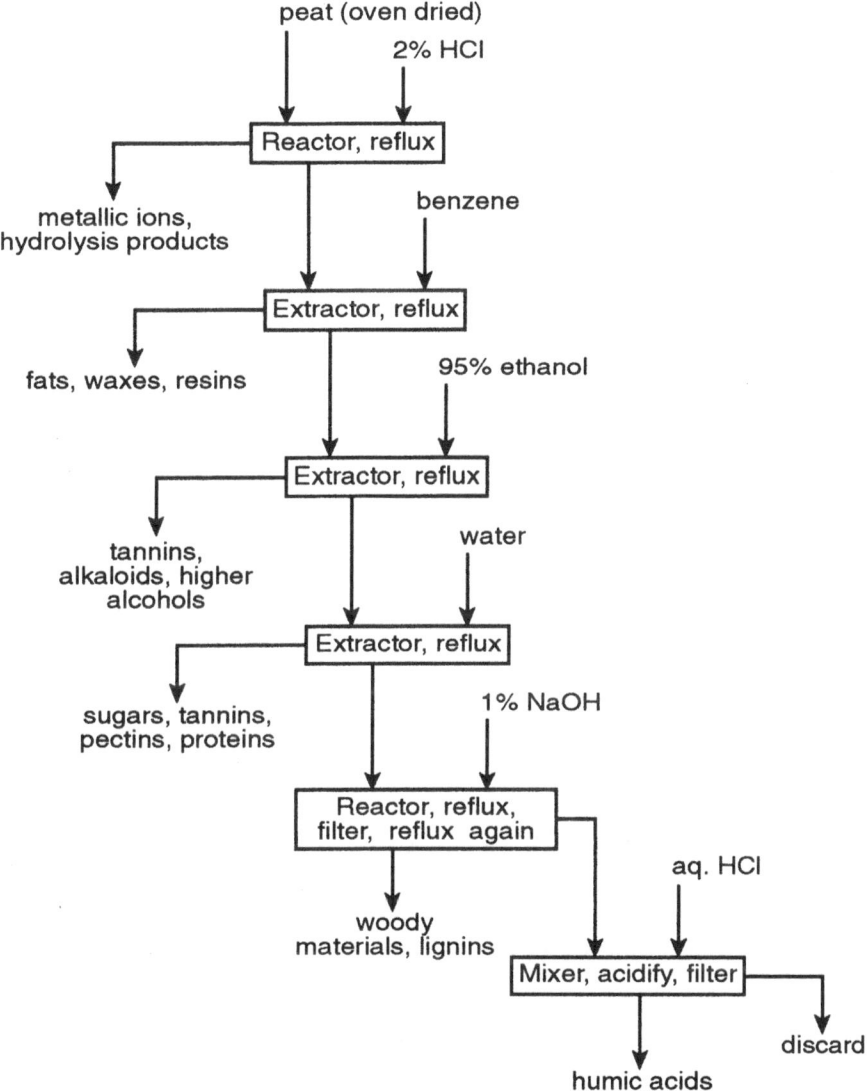

Figure 2.1 *A separation scheme for isolation of peat humic acids*

(Source: Arnold, Lowy and Thiessen, 1925)

Chemical products from peat

The amount and type of chemical feedstocks available from peat are
dependent upon several factors, including the plants which comprise the
peat, the source and impurity of the associated water table and the altitude
of the peatland. According to Fuchsman (1980) in his comprehensive
study on this subject, four major groups of substances may be produced
from peat and include:

- peat waxes, resins and asphalts;
- sugars;
- metallurgical coke and activated carbon; and
- humic acids.

Peat waxes Peat waxes, resins and asphalts are derived from the bitumens
in the peat which are extracted by combining a solvent such as petroleum
ether or benzene with peat. The bitumens will dissolve and are recovered
when the solvent is evaporated. Peat waxes may be used in the manufacture
of shoe polish, furniture polish, paper sizing, cosmetics, salves and
ointments.

Sugars and other carbohydrates Sugar suitable for the production of
yeast can be produced from hydrolysis of peat by an acid such as sulphuric
acid. Traditional yeasts such as baker's and brewer's yeasts and high-
protein yeasts for animal feed can be produced using these sugars.
Organic chemicals such as furfural, which can be used in the manufacture
of plastics, and lactic and glycolic acids can also result when peat
undergoes hydrolysis.

Coke and activated carbon The pyrolysis of peat leads to the production
of carbon. Semi coke, which is used as an industrial fuel, will be formed
during the early stages of pyrolysis. Activated carbon, an adsorbent, can
also be used to remove impurities from starch and sugar and for water
purification and sewage treatment.

Humic acids Approximately two-thirds of the residue from the sulphuric
acid hydrolysis of peat will result in humic substances from which humic
acids can be derived. These humic acids, when used as fertilizer, can
stimulate plant growth to increase plant yields. The breakdown of peat to
obtain chemical feedstocks is much more complicated than outlined here.
Figure 2.1 gives some indication of the number of steps required to obtain
humic acids.

Other products from peat Peat contains a growth inhibitor which in the former Soviet Union is used to produce Torfat. Torfat is added to an ointment and used to treat eye infections and disorders such as progressive myopia (short-sightedness). Peat tar, which can result from peat pyrolysis or be removed from gasified peat, contains a number of phenols which may be extracted to aid in the production of adhesives. Acetic acid may also be produced from phenolic waters.

Chemical feedstocks from lignite

Over a century ago coal was the primary source for many organic chemicals. Gradually, as petroleum and natural gas became more widely available, the use of coal to produce chemical feedstocks diminished. A vast number of products are produced using organic chemicals, most of which may be derived from coal, including lignite.

Today, petroleum and natural gas remain the preferred raw materials for petrochemical production. However, for political reasons (e.g. South Africa), or because it is a readily available domestic fuel (Germany), the manufacture of chemical feedstocks from coal is still an important industry.

A great deal of research has been done on coal chemistry. A brief overview is given here to show the potential non-fuel uses of lignite and to point out the environmental issues surrounding lignite as a source of chemical feedstocks. Using lignite instead of other coals reduces the cost of the raw material for petrochemical manufacture. This cost advantage, however, must be weighed against the higher costs incurred for the additional processing which lignite requires such as drying.

In some cases, lignite may be preferred to high-grade bituminous coals because of a greater reactivity. This is true in certain gasification processes.

Gasification and liquefaction are the two major processes (see Chapter 1) which begin the breakdown of coal into chemical feedstocks. Pyrolysis can also be used, but in practice it is much less common. Figure 2.2 illustrates the basic groups of chemicals which may be produced from coal.

By-products produced from lignite include aromatics, olefins, methanol and other alcohols. The most common aromatics produced from coal include hydrocarbon oils such as benzene, toluene and xylene. Derivatives from these substances include benzol (a petrol additive which prevents engine knocking), polyurethane foams and microscopy aids such as cleaning agents and tissue-mounting material.

Derivations of olefins such as ethylene, propylene and butadiene are used to manufacture synthetic rubber, nitroglycerin, printing inks and carpet backing. In the food industry ethene is used to mature fruit in

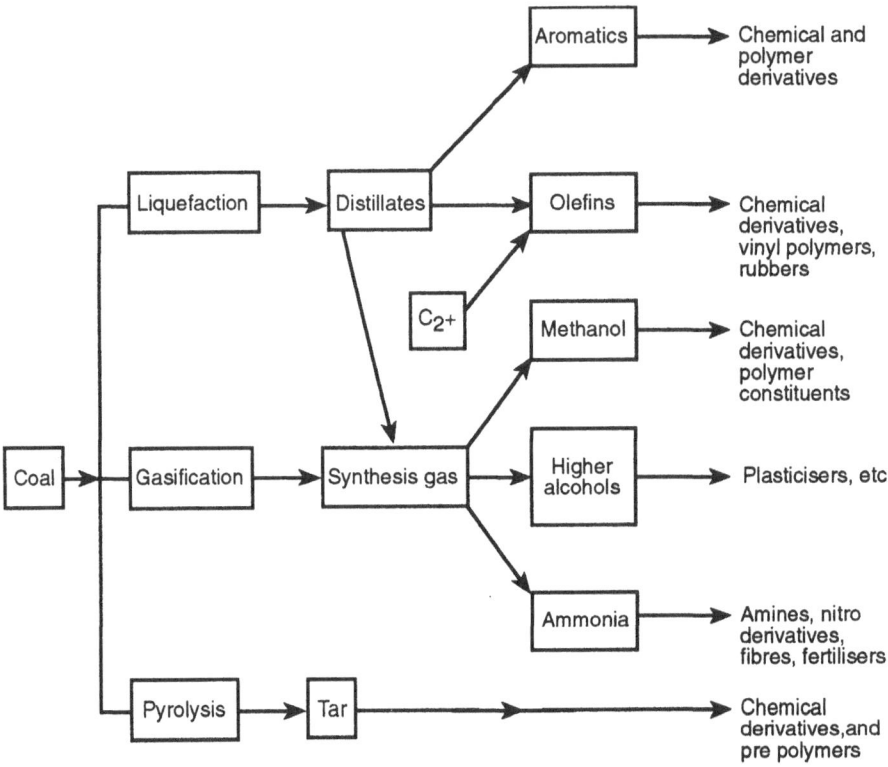

Figure 2.2 *Processes that currently provide chemicals and polymers from coal*

(Source: Adapted from Davies, 1985)

storage, and glycerin is used in the manufacture of food products such as icing.

Methanol is an alcohol which has many derivatives. It is a commonly used solvent which is also known as methyl alcohol. Higher alcohols are alcohols which only boil at a very high temperature and are also a potential lignite by-product. They are used in the manufacture of plasticizers for lacquer and certain types of plastics (e.g. polyvinyl chloride).

Chemical feedstocks from oil shale

Kerogen is a chemically complicated substance which accounts for the largest proportion of organic material in oil shale. The chemical composition of kerogen and the chemical reactions of oil shale are not fully understood (Lee, 1991; Aarna, 1978).

This lack of knowledge concerning oil shale is in part due to the use of crude oil chemical analysis methods to determine the properties and characteristics of oil shale. The relevancy of these methods is sometimes uncertain. In addition, the development of oil shale resources is not widespread, so oil shale research is limited in comparison with other more popular fuels such as oil, natural gas and coal. The great variation in mineral matter and organic content also hinders major breakthroughs in this field.

Figure 2.3 illustrates some of the major products which may be produced from oil shale. The current level of knowledge regarding the chemical structure of oil shale is generally limited to the groups of chemical compounds present in kerogen. The majority of individual chemicals and simple chemical compounds remains a mystery. As research continues, however, further applications and products from oil shale-derived chemical feedstocks may come to light.

Phenols The largest known group of chemical compounds in oil shale is phenols, which constitute up to 30 per cent of shale oil and are also present in water used in oil shale processing. Included in the phenol group is resorcinol, which is used in the production of resins such as epoxy. When mixed with formaldehyde, phenols can be used in the preparation of cold-setting adhesives.

Organic acids Kerogen can be oxidized to produce organic acids, which have a great number of applications in industry. One important organic acid is adipinic acid, which is used in the production of synthetic fibre. Other acids which may be oxidized from kerogen include succinic acids, used in the manufacture of alkyd resins; sebacic acid, from which esters used in the production of resins and frost-resistant plasticizers are derived; and other dibasic acids such as pimelic acid and suberic acid.

Neutral oxygen compounds Neutral oxygen compounds also represent a major component of shale oil. Appropriate applications for these compounds have not yet been determined. In the former Soviet Union neutral oxygen compounds have been added to lubricants and diesel fuel in experiments to increase the performance of these substances.

Activated carbons Recent research has increased the ability to produce adsorbent or activated carbons, carbon fibres and activated carbon fibres from oil shale. These materials have a large range of applications, including

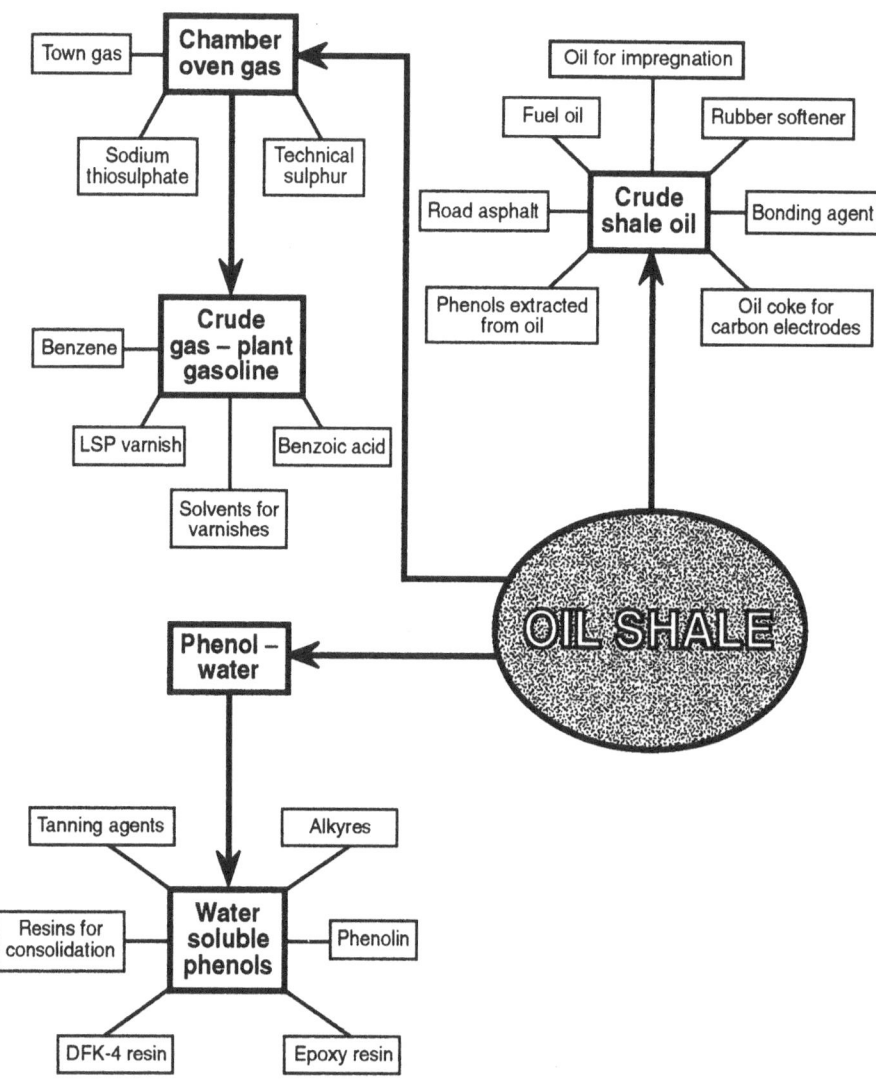

Figure 2.3 *Products from oil shale*

(Source: Valgus, 1974)

wastewater treatment, respirators, solvent recovery and air and gas purification (Derbyshire et al., 1995).

Chemical products from wood

The distillation or pyrolysis of wood produces a number of by-products, some of which are chemical feedstocks. These include methyl alcohol, acetic acid, naphthalenes and ethyl alcohol. It is commonly used as an industrial solvent. Methyl alcohol is also known as methanol or wood alcohol. A major pyroligneous acid is acetic acid, which plays an important role in the textile industry. Acetic acid is also used for pickling food both on the industrial and household levels. Acetone is produced using this acid. A product of wood tar, naphthalene is a chemical compound, derivatives of which are important in the manufacture of dyestuffs.

Ethyl alcohol, also known as ethanol, can be produced from the cellulosic biomass in wood, herbaceous and woody plants, agricultural residue and MSW. Ethanol neat or blended may be used as a petroleum substitute to power transportation vehicles. It may also be used to produce drinking alcohol such as vodka.

The environmental impacts associated with the production of chemical feedstocks from low-grade fuels

The potential negative environmental impacts resulting from the manufacture of chemical feedstocks from low-grade fuels are similar to impacts of the fuel cycle, since chemical feedstock production also requires mining, processing, transport, conversion and utilization. The magnitude of these impacts is also determined by the scale of operation and local environmental standards and regulations.

Typically, the production of chemical feedstocks requires less mining and transport as compared to energy production as smaller volumes are usually involved, although greater impacts may occur during the processing/manufacturing stage where solvents and adsorbents are added. The addition of these substances may result in the formation of chemical compounds which should be prevented from escaping into the surrounding environment.

Many processes for the production of chemicals use large volumes of water. The acquisition of this water may present environmental problems, while the disposal of large quantities of polluted water is an obvious difficulty. In most modern process cycles there is an attempt to operate a closed system with respect to water use, although this may not always be possible. This wastewater can, in effect, form a toxic waste stream filled with potentially dangerous containments such as organics, and its safe disposal must be ensured.

The possible environmental effects of chemical feedstock production from low-grade fuels along with proposed measures for elimination or reduction of these effects are summarized in the box at the end of each topic below. Overall, it should be emphasized that the cost or complication of ensuring environmental protection will be greatly reduced if the appropriate measures are part of the original design of the chemical feedstock production process.

Mining or gaining for chemical feedstock production

The environmental consequences of mining to obtain raw material for chemical feedstock production are similar to those impacts described and assessed in Chapter 4 on mining or gaining for fuel purposes (Table 2.1). There are currently very few incidents of the mining of low-grade fuels specifically for chemical feedstock production. A notable exception to this is oil shale in the former Soviet Union.The importance of oil shale, lignite, wood, municipal wastes and peat as sources of chemical products is greatly overshadowed by oil and the higher-grade coals.

The disruption to the existing landscape and ground cover is usually minimal when compared to mining or gaining for fuel purposes, since smaller volumes of fuel are removed than for industrial energy or power production. This, in turn, means that there is a much smaller loss of wildlife habitat and less likelihood of the loss of any rare or endangered plant species. There is also less chance of a significant disturbance of the accompanying water table and surface waters.

The small scale of mining to provide chemical feedstocks allows for easier storage of valuable topsoil and subsoil with lignite and oil shale mining and small labour and raw materials costs for revegetation and general restoration of the area.

Transportation

The movement of vehicles to and from a chemical production plant may disturb residents in the vicinity of the plant depending upon the scale of operation and existing background traffic levels. There will also be an increased risk of accidental spillage of the various chemical products along access routes. An upset or spill of highly toxic chemicals such as phenols can pose a severe threat to the environment and to public health.

Table 2.1 *Potential impacts and control measures during mining for chemical feedstock production*

Potential impact	Environmental protection measure
■Land use disruption	■Planning for land restoration prior to project initiation
■Loss of surface vegetation	■Revegetation with regionally appropriate species
■Loss of soil fertility	■Storage of topsoil during mining; reapplication during restoration; addition of soil amendments
■Loss of wildlife habitat	■Survey for wildlife living at proposed site; mining for chemical production purposes requires small area, so move proposed site to avoid habitat loss or disruption

The choice of transport route and the method of transport must be carefully considered. This risk may be reduced through careful training of the drivers of chemical transport trucks, and by assessing local traffic patterns to determine when and where traffic is lightest and scheduling transport to take place during these times and along these routes. Route selection must ensure the avoidance of sensitive community features such as schools, and transport of chemical feedstocks or waste from the plant by either truck or rail must take place in vehicles that have been proven to withstand high impact or pressure. These potential impacts are summarized in Table 2.2.

Heating and cooling

The production of chemicals almost always requires steps involving heating and cooling, both of which may give rise to undesirable effects (Table 2.3) (Fuchsman, 1980). Heating results from combustion which, in turn, can lead to emissions of sulphur, NO_x, ash and substances which are highly volatile such as chlorine and fluorine, depending upon the composition of the fuel. Some or all of the heat generated during combustion may be used at the plant for space heating.

Table 2.2 *Potential impacts and control measures during transportation of chemical feedstocks*

Potential impact	Environmental protection measure
■Accidental spill of hazardous chemical products	■Special driver training; choose route and travel times to minimize risk of traffic accident along transport routes; transport in secure trucks or rail cars able to withstand the potential impact and pressure of a collision or accident; do not transport hazardous waste past sensitive community features such as schools or built-up residential areas

Table 2.3 *Potential impacts and control measures for heating and cooling during chemical feedstock production*

Potential impact	Environmental protection measure
Heating	
■Undesirable emissions (e.g. sulphur, NO_x, chlorine, ash)	■Installation of appropriate control mechanisms at plant (e.g. electrostatic precipitators, addition of lime)
Cooling	
■Production of effluent with temperature higher than the surrounding environment	■Allow effluent to cool in settling ponds or through addition of cool water prior to release

Cooling from production processes creates warm effluent (often water) even after internal re-use of most of the usable heat content. This heated liquid must be combined with enough cool water to ensure that the environmental impacts of the discharged effluents are minimal. For example, warm water may interfere with the successful reproduction of certain species of fish as has been demonstrated in work done on rivers adjacent to nuclear power plants (Christie, 1988).

Extraction

The use of certain solvents and purifying agents in the production of chemical feedstocks can result in effluent and vapours containing harmful substances (Table 2.4). The application of chromic acid to manufacture peat wax can lead to the presence of chromium in the effluent which will contaminate soil and water if it is not recovered or recycled within the plant.

Solvent vapours, particularly those containing sulphur or chlorine, must be carefully handled to avoid a negative impact on air quality. The combustion of these vapours at an appropriate temperature, with the addition of limestone, can capture and control the escape of sulphur and chlorine (Müncher and Schilling, 1985). Electrostatic precipitators (ESPs) are also effective in controlling these types of emissions.

Table 2.4 *Potential impacts and control measures during extraction of chemical feedstocks*

Potential impact	Environmental protection measure
■Production of effluents and vapours causing contamination of groundwater, adjacent surface waters and soils	■Operate using a closed system whenever possible; closely monitor all potentially hazardous substances. Install efficient solvent recovery systems; solvent vapours (after removal of sulphur and chlorine) may be burned; reducing agents, adsorbents and settling ponds may be used to isolate, stabilize and/or neutralize undesirable chemical compounds

Wastewater treatment

Nearly all chemical feedstock plants produce large volumes of contaminated water effluents. These may range from lightly contaminated cooling water to grossly contaminated process effluent. Any modern feedstock plant, whether based on coal or other raw materials, must contain elaborate treatment facilities to eliminate these contaminants.

The technology of wastewater treatment is reasonably well developed, and there are several standard texts on the subject. In general, modern technology involves separating contaminated liquids into two streams: a stream of relatively pure water which can be recycled into one of the water circuits, and another stream comprising concentrated chemicals which can be incinerated (sometimes with a positive heat value) or even sold as a chemical by-product.

In the event that this approach is not possible, a residue of highly toxic waste may remain. The transportation of this hazardous waste from the site and its disposal is an important factor in the environmental safety of the plant. Historically, a great deal of contaminated land and toxic waste damage has resulted from coal conversion processes in Europe and North America. Modern production of chemical feedstocks and other fuel processing must totally elimate both of these impacts.

Fertilizers and soil amendments from low-grade fuels

Sphagnum peat moss and peat-derived fertilizers

In most countries peat is more valued for its contribution to agricultural and horticultural practices than for fuel purposes. *Sphagnum* peat moss is available in a number of different shapes and sizes (e.g. peat pots, peat pellets, peat bales) and provides a suitable growing medium for many types of seeds. Peat used in this way is characterized by light weight and high water holding ability. As indicated earlier in this chapter, peat is also the raw material from which humic acid may be derived. Humic acid can be applied as a plant growth stimulant and is used as a fertilizer in many parts of the world.

Although in many regions of the globe peat forms an abundant resource which can be utilized without serious impacts on existing ecosystems, there are some places where horticultural peat can have negative ecological consequences. This is where relatively limited peat reserves are located close to major horticultural and gardening markets and are, therefore, developed beyond the ability of the characteristic ecology to be sustained. Examples of this occur in the low-level peat moors of England.

Alternatives to peat include low-peat and no-peat composts. Low-peat composts (where the peat content is less than 20 per cent) may comprise peat and materials such as coconut fibre, sand, perlite, bark and treated farm waste. Peatless composts include coconut fibre, graded bark mix and vermiculite.

Nerosine

Nerosine, made almost entirely from shale oil, is used in the former Soviet Union, principally in Estonia, to combat erosion by 'fixing' fine soils and sands. Nerosine is sprayed on the soil at the rate of 1–1.5 tonnes per hectare. This facilitates oxidation and the nerosine becomes humuslike, binding the fine soil particles together (Aarna, 1978).

There is some evidence, however, to indicate that the use of nerosine contributes to an increase in levels of PAHs in the soil. As Table 2.5 illustrates, benzo(a)pyrene, a known carcinogen, has been found in much greater quantities in soils treated with nerosine than in those where nerosine was not applied.

Table 2.5 *BaP content (µg/kg) in soil samples treated with nerosine in May 1969 in (a) Uzbek SSR (cotton) and (b) Kazak SSR (spring wheat)*

(a)

Soil depth (cm)	Control		Nerosine 1.4 t/ha		Nerosine 2.1 t/ha	
	May	September	May	September	May	September
0–10	4.7	8.2	64.0	47.0	235.0	153.0
10–20	4.9	2.8	15.0	47.0	46.0	46.0
20–30	4.4	5.0	32.0	54.0	9.0	79.0

(b)

Soil depth (cm)	Control				Nerosine-treated			
	1969			1970	1969			1970
	June	July	August	August	June	July	August	August
0–10	0.80	1.0	3.5	1.0	238.0	186.0	169.0	37.0
10–20	0.34	1.0	1.2	0.8	95.0	42.0	14.0	7.2
20–30	0.10	0.3	1.2	0.1	9.0	25.0	21.0	5.2

(Source: Shabad, 1973)

Liming with powdered oil shale ash

Powdered oil shale contains calcium compounds which, when added to acidic soil, will help to increase pH values. An experiment in Estonia in a forest ecosystem where the predominant species was *Pinus sylvestris*, and where the extent of root rot and scleroderis canker provided evidence that the system was under stress, showed that the application of powdered oil shale ash was highly effective in reducing the acidity of the forest soil and thus in providing improved conditions for forest growth (Terasmaa, 1994).

Gypsum

Wet scrubbing is the most commonly used flue gas desulphurization (FGD) technique. During wet scrubbing, limestone in an aqueous medium is used to react with SO_2 gas. SO_2 is one component of coal (or oil) combustion. The chemical reaction gives calcium sulphate, or gypsum. Incomplete reaction can also lead to the formation of sulphite salt. More than 95 per cent of the SO_2 emitted may be abated in this way. Gypsum, which occurs naturally, is thus a by-product of the FGD process. Its quality is variable depending upon the balance of sulphite and sulphate formed and the degree of dewatering necessary. Gypsum may be used as a soil additive for sulphur-deficient soil or to correct cation imbalances. It may also be used as a soil physical conditioner.

Absorbent biofilters

The use of peat in specially designed filters to purify contaminated wastewater from industrial processes is becoming more common (Allen, 1994; Mota, 1994). In one study peat was found to be effective in absorbing mercury from wastewater within a period of five hours (Virarghaven, 1995). *Sphagnum* peat has also been used in biofilters that act as an alternative to conventional tile beds associated with septic tanks, and for cleaning polluted water. In field trials the peat filter removed 97.8 per cent biological oxygen demand (BOD), 96.1 per cent of total suspended solids and 99.5 per cent of fecal coliform bacteria at a temperature of 12 to 16°C (Jowett, 1995). Other experiments indicate the effectiveness of brown coal, in addition to peat, for removing heavy metals from wastewater (Horacek, 1994).

Building materials

During efficient combustion of low-grade fuels most of the organic matter in the fuel is burned. The inorganic matter, the mineral component of the fuel, is left in the form of ash. This is especially true for lignite and oil shale since peat generally has a higher proportion of organic material than lignite or oil shale. Although the contents of ash will vary greatly in general, the ash from oil shale and lignite pulverized fuel combustion can be used in cement-making. Depending upon its composition, the ash may also be used in the manufacture of brick, as an asphalt filler in road construction and also as hard core. Ash from fluidized bed systems is still in limited use, although research is being conducted on the suitability of this ash for the production of lightweight bricks, gas concrete as well as road construction materials (Hein, 1990).

Gypsum may also be used in the manufacture of materials used principally for building. It is important in the making of plaster of Paris and plaster-board as well as wall linings, light bricks and floorboards.

Landfill

Depending upon its composition and properties, mining spoil and ash from low-grade fuel combustion may be used for landfill or in land reclamation schemes. Ash material from lignite (and oil shale), although alkaline in reaction, has been used as an amendment for spoil materials in reclamation work, particularly for acid colliery spoil material. Ash from low-grade coal has also been employed as landfill material for other excavation sites (clay pits and quarries, for example). Lignite mine spoil, spent oil shale and oil shale ash are difficult to wet, may be saline, suffer from ion imbalance and have low fertility. These features need correction. A number of studies have investigated these problems and identified methods of amelioration (Chadwick 1992; Bradshaw and Chadwick, 1980; Knabe, 1973).

References

AARNA, A (1978) *Chemical Engineering in the Estonian SSR*. Perioodika, Tallinn.

ALLEN, S J (1994) 'Peat as an absorbent for dyestuffs and metals in wastewater'. *Resources, Conservation and Recycling* **11**: 25–39.

ARNOLD, C L, LOWY, A and THIESSEN, R (1925) 'Humic acids from peat'. *Fuel* **14**: 107–112.

BRADSHAW, A D and CHADWICK, M J (1980) *The Restoration of Land*. Blackwell, Oxford.

CHADWICK, M J (1992) 'Environmental implications of the use of low-grade fuels'. In *The Environmentally Sound Management of Low-grade Fuels*. SEI, Stockholm.

CHRISTIE, G J (1988) 'Unpublished Masters Thesis'. Institute for Environmental Studies, Toronto.

DAVIES, G O (1985) 'Chemical feedstocks by the direct liquefaction of coal'. In *Chemicals from Coal: New Developments*. Blackwell Scientific Publications, Oxford.

DERBYSHIRE, F, GRAHAM, U, QING FEI, Y, ROBL, T, JAGTOYEN, M (1995) 'Oil shale as a feedstock for carbon materials'. In *Composition, Geochemistry and Conversion of Oil Shales*. Kluwer Academic Press, Dordrecht.

ELLIOT, M A (1981) *Chemistry of Coal Utilization.* Vol 2. John Wiley, New York.

FALBE, J (1982) *Chemical Feedstocks from Coal.* John Wiley, New York.

FUCHSMAN, C H (1980) *Peat: Industrial Chemistry and Technology.* Academic Press, New York.

HAMERSLAG, F E (1985) *The Technology and Chemistry of Alkaloids.* D. van Nostrand, New York.

HEIN, K (1990) 'Byproduct from combustion systems'. *VTT Symposium 108: Low-grade Fuels.* Technical Research Centre of Finland, Espoo.

HORACEK, J (1994) 'Purification of wastewaters containing low concentration of heavy metals'. *Journal of Hazardous Matter* **37**: 69–76.

JOWETT, E C (1995) 'On-site wastewater treatment using unsaturated absorbent biofilters'. *Journal of Environmental Quality* **24**: 86–94.

KLAR, M (1925) *The Technology of Wood Distillation.* Chapman and Hall, London.

KNABE, W (1973) 'Development and application of the Domsdorf ameliorative treatment on toxic spoil banks and lignite opencast mines in Germany'. In *Ecology and Reclamation of Disturbed Land* (ed by R J Hutnik and G Davis). Gordon and Breach, New York.

LEE, S (1991) *Oil Shale Technology.* CRC Press, Florida.

LOWRY, H H (1945) *Chemistry of Coal Utilization.* Vols 1 and 2 and suppl vol (1963). John Wiley, New York.

MOTA, A M (1994) 'Adsorption of humic acid on a mercury/aqueous solution interface'. *Water Resources* **28**: 1285–1296.

MÜNCHER, H and SCHILLING, H (1985) 'Fluorine and chlorine emissions from FBC enrichments in flue ash and filter dust'. In *Proceedings of the 8th International Conference on Fluidized-bed Combustion.* Vol 3. Houston, Texas.

PAYNE, K R (ed) (1987) *Chemicals from Coal: New Processes.* John Wiley, Chichester.

PAYNE, K R (ed) (1985) *Chemicals from Coal: New Developments.* Blackwell Scientific Publications, Oxford.

PERRY, J H (1963) *Chemical Engineers Handbook.* 4th edition. McGraw-Hill, New York.

SHABAD, L M (1973) *Circulation of Carcinogens in the Environment.* Meditsina, Moscow.

SHREVE, R N (1967) *Chemical Process Industries.* 3rd edition. McGraw-Hill, New York.

STRATTON, A (1983) 'Energy and feedstocks in the chemistry industry'. *Proceedings of International Conference* (Brussels, 1982). Ellis Harwood, Chichester.

TERASMAA, T (1994) 'Liming with powdered oil shale ash in a heavily damaged forest ecosystem'. *Proceedings of the Estonian Academy of Science* **4**: 101–108.

VALGUS (1974) *The V I Lenin Oil Shale Processing Complex*. Valgus Publishing House, Tallinn.

VIRARGHAVEN, T (1995) 'Adsorption of mercury from wastewater using peat'. *Journal of Environmental Science and Health* **30**: 553–566.

3

Global resources and reserves of low-grade fuels

Throughout the world, exploration and inventory have identified quantities of low-grade fuels. The exact amount of any particular fuel is not known with certainty. Some of these fuels may be very difficult to harvest or mine with current technology but, through future innovations, may one day be retrievable. All known and unknown quantities of a fuel, whether recoverable or not, are known as resources. The resources which are currently obtainable are known as reserves.

Information on energy resources and reserves is notoriously undependable with respect to consistency of reporting among countries. The World Energy Conference (WEC) 1989 *Survey of Energy Resources* represents the most up-to-date data on reserves and resources of most low-grade fuels at the time of writing. Although all countries which contributed to the survey were encouraged to collect and present their data in a uniform manner, instructions are always open to a variety of interpretations, and hidden differences in energy accounting may occur. For example, some countries have completed a nationwide survey of low-grade coal resources, while others are simply relating information on currently operating sites.

If a fuel is not traded internationally and is primarily used as a low-cost or free source of energy for household purposes, the quantities consumed are not easily determined. There is no 'visible trade' in these energy sources, but the importance of these fuels, almost all of which are low-grade, should not be underestimated. They provide an increased level of comfort and health which might otherwise be denied to those who cannot afford commercial fuels or metered sources of energy.

The information given in this chapter is an attempt to give a broad overview of known quantities and global distributions of low-grade fuels. It should be viewed as a general estimate, keeping in mind that energy use patterns are constantly changing and that precise energy data for many countries and regions are simply not available.

Low-grade coals

Low-grade coals are generally high-moisture or high-ash coals, some of which have a high sulphur content. Most low-grade coals are lignite.

During mining and processing of higher-grade coals, low-grade coals may be created (Couch, 1989).

A 1989 survey by Couch, summarized in a report by the Low-grade Coal Committee of the World Energy Conference, collected information on deposits of low-grade coal in countries generally known to possess significant quantities. Couch (1989 and 1988) and Christov and Todoriev (1989) defined low-grade coal as coal with a calorific value less than 16 MJ/kg (approximately 4000 kcal/kg) as mined. This differs only slightly from the working definition outlined in Chapter 1 (17.5 MJ/kg, as mined, plus fuels with a high impurity content). The data are presented in Table 3.1.

Australia, Bulgaria, Canada, China, Czechoslovakia, Germany, Greece, Hungary, India, Korea, Poland, Spain, Turkey, the United States, the former Soviet Union and the former Yugoslavia have been identified as the major low-grade coal-producing countries. High moisture coals are found among the low-grade coals in Australia and Germany. Turkey, Korea and Brazil have deposits of particularly high ash coals (Couch, 1989).

Figure 3.1 summarizes information on the global distribution of major sedimentary deposits which represent the areas where major lignite deposits are most likely to occur.

Peat

Peat is primarily used in agriculture as a growing medium or a soil amendment. It is an important addition to the national energy balance in Ireland, Finland and parts of the former Soviet Union. Peat exists in various forms (see Chapter 1) almost all of which are burnable after drying, although the low level of plant decomposition in *Sphagnum* peat makes it highly undesirable for combustion.

There is little trading of fuel peat in the international marketplace, but in poorer regions and countries where it is available as a 'free' good it will, like wood, be used to provide heat for warming the home and for cooking. No figures are available for the quantity of peat used in this way.

Table 3.2 presents information on known areas of peatland, while Table 3.3 gives details on how those resources are used. Figure 3.2 provides an illustration of the distribution of peatland suitable for fuel.

Oil shale

The world's reserves of oil shale are much greater than those known for crude oil (Snape, 1995). Oil shale deposits tend to be vast; for example,

Table 3.1 *Low-grade coal deposits*

Country	Total coal in place Mt	Total recoverable reserves Mt	Coal production in 1988 Mt/yr	Share of potential recoverable reserves %
Australia	48233	38691	47.7	16.1
Brazil	15788	5999	18.6	77.1
Bulgaria	4197	3063	33.5	18.4
Canada	3025	2217	10.0	54.2
Czechoslovakia	10475	8024	104.3	38.3
Denmark	63	–	–	–
FRG	–	4150	109.4	38.3
GDR	–	40000	300.0	–
Greece	5312	3000	34.0	–
Hungary	3603	264	18.0	79.9
India	93000	17000	100.0	–
Indonesia	18000	–	–	–
Japan	547	49	–	–
Korea	132	90	32.5	–
New Zealand	11701	7028	0.2	64.8
Poland	13200	–	73.0	–
Portugal	41	36	0.3	91.7
South Africa	1783	1104	2.0	90.1
Spain	2096	963	21.5	36.9
Thailand	1705	657	5.9	3.5
Turkey	–	874	20.1	14.4
USA	40886	32709	67.0	–
USSR	105000	94500	122.0	–
Yugoslavia	16000	15000	65.0	–
Zambia	68	44	0.5	–

(Source: Christov and Todoriev, 1989)

in the former Soviet Union oil shale can be found over 11,300 square kilometres. Oil shale in the former Soviet Union is used for the production of fuel and is an important source of chemical feedstocks. In the United States and South Africa oil shale contributes in a small way to the national production of liquid fuels through the extraction of shale oil. The use of oil shale is not widespread, particularly because the cost of mining and retorting oil shale is far from competitive with oil, natural gas and coal.

Figure 3.1 *Global distribution of major sedimentary deposits with which major lignite deposits are likely to be associated*

● Sedimentary basins

(Source: Smith, 1981)

Table 3.2 *Areas of peatland*

Area given in 1000s of hectares			
Angola	*	Iceland	1000.0
Argentina[a]	45.0	India	32.0
Australia[b]	15.0	Indonesia	17000.0
Austria	22.0	Ireland	1180.0
Bangladesh	60.0	Israel	5.0
Belgium	18.0	Italy	120.0
Belize	68.0	Jamaica	21.0
Bolivia	0.9	Japan	250.0
Brazil	1500.0	Korea (North)	136.0
Bulgaria	1.0	Liberia	40.0
Burundi	1.0	Luxemburg	0.2
Canada	150000.0	Madagascar	197.0
Chile	1047.0	Malawi	91.0
China	4200.0	Malaysia	2500.0
Colombia	339.0	Mozambique	*
Congo	290.0	Netherlands	280.0
Costa Rica	37.0	New Zealand	150.0
Côte d'Ivoire	32.0	Nicaragua	371.0
Cuba	767.0	Norway	3000.0
Czechoslovakia	31.0	Panama	787.0
Denmark	120.0	Papua New Guinea	*
El Salvador	9.0	Philippines	6.0
Falkland Islands	1151.0	Poland	1300.0
Fiji	4.0	Puerto Rico	10.0
Finland	10400.0	Romania	7.0
France	90.0	Rwanda	80.0
French Guiana	162.0	Senegal	1.5
FRG	1110.0	Spain	6.0
GDR	489.0	Sri Lanka	2.5
Greece	5.0	Surinam	113.0
Guinea	525.0	Sweden	7000.0
Guyana	814.0	Switzerland	55.0
Honduras	453.0	Thailand	68.0
Hungary	30.0	Trinidad and Tobago	1.0

Table 3.2 (continued)

Area given in 1000s of hectares

Uganda	1420.0	Venezuela	1 000.0
United Kingdom[c]	1580.0	Vietnam	183.0
Uruguay	3.0	Yugoslavia	100.0
USA[d]	59640.0	Zaire	*
USSR	150000.0	Zambia	1106.0

* Extent of peatland not assessed
[a] Refers to peat from Tierra del Fuego
[b] Data for Australia refers to Queensland
[c] Deposits of peat vary considerably in size; most peat found in Scotland and Northern Ireland
[b] Data for USA includes Alaska with a peatland area of 49,400 kha

(Source: Bord na Móna, 1985)

Table 3.3 *Peat resources and use worldwide*

Country	Resource dry weight Mt	Use, Mt/yr (40% H_2O)	
		Fuel	Horticulture
Canada	600000	0.00	0.50
China	12500	0.80	1.30
Finland	41600	3.10	0.50
France	<1000	0.05	0.10
FRG	4500	0.25	2.00
Iceland	4000	–	–
Indonesia	68000	–	–
Ireland	4700	5.57	0.38
Japan	1000	–	–
Malaysia	10000	–	–
Netherlands	1120	–	–
Norway	12000	0.00	0.08
South America	24700	n/a	n/a
Sweden	28000	–	0.27
United Kingdom	6300	–	–
USA	238560	0.00	0.80
USSR	n/a	80.00	120.00

(Source: Spedding, 1988)

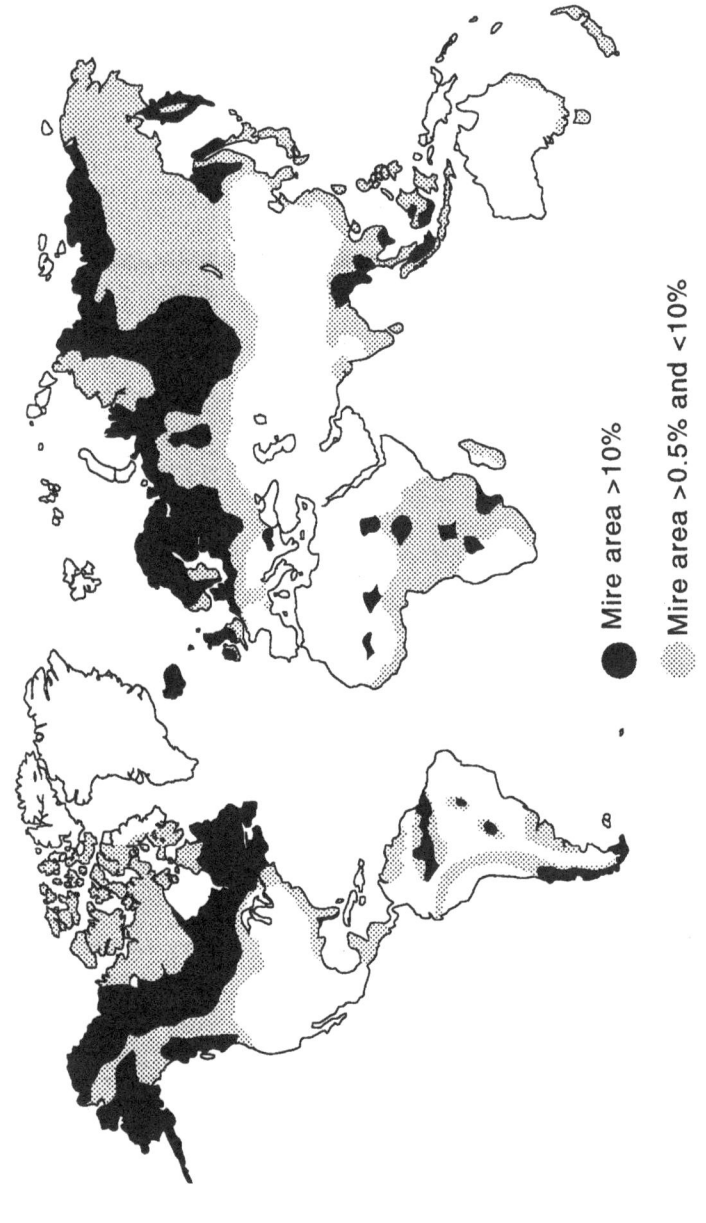

● Mire area >10%

Mire area >0.5% and <10%

O Mire area <0.5%

Figure 3.2 *Global distribution of peat*

(Source: Gore, 1990)

Table 3.4 provides details on known resources and reserves of oil shale. Figure 3.3 presents information on the general location of oil shale deposits around the world.

Table 3.4 *Resources and reserves of oil shale at the end of 1987*

	Recovery method (see foot-note)	Proved amount in place million tonnes	Proved recoverable reserves million tonnes	Deposit area (km²)
Australia	M	–	–	–
Brazil	R	965	352	870.0
China[a]	R	–	–	–
Estonia	–	6546	2000	11300.0
Israel	M	11000	700	180.0
Jordan	M	40000	4000	650.0
Morocco	–	–	1600	–
New Zealand	–	–	–	3.3
South Africa	P	–	–	–
Thailand	P	18000	807	53.0
Turkey	–	842	227	–
Uruguay	M	825	40	126.0
USA	R	217000	–	–
USSR[b]	–	260	–	–

– Unknown or zero

Recovery method: M - surface mined
P - in situ recovery
R - retorting

[a] Only figures available are calculated on oil from shale, which is 10⁹ barrels
[b] Beloselsky, 1991

(Source: Adapted from WEC, 1989)

Wood

To varying degrees of density, wood grows in every country in the world. Table 3.5 illustrates that wood is used to produce energy in both developed and developing countries.

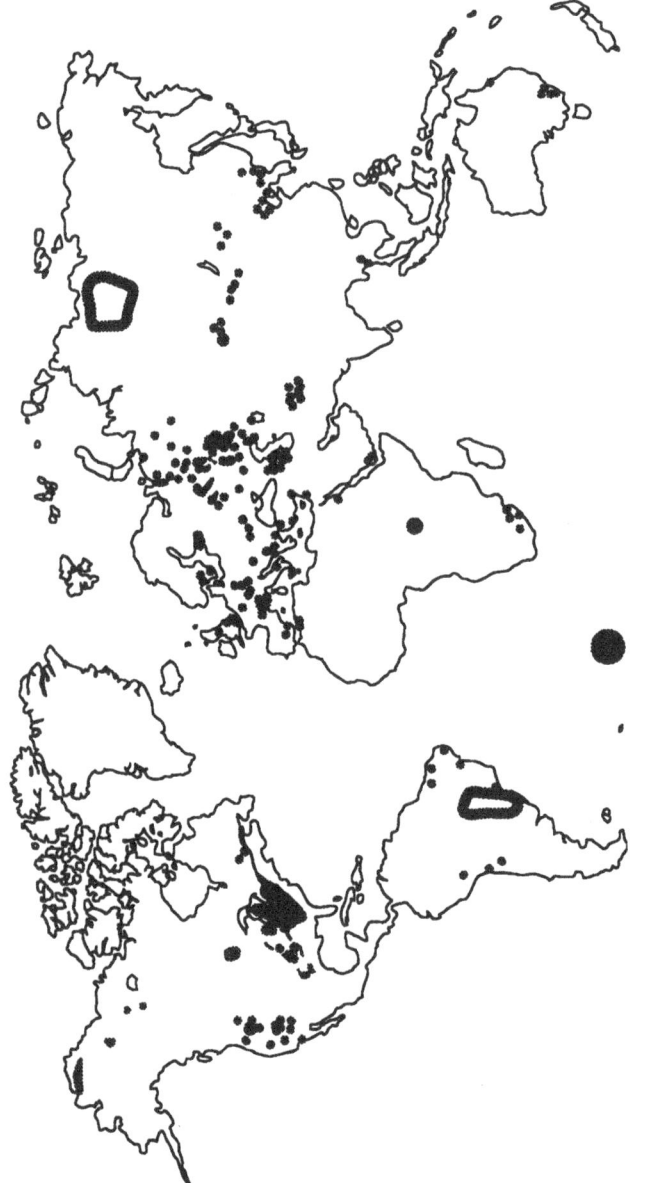

Figure 3.3 *Global distribution of major oil shale deposits*

(Source: Compiled from Yen and Chilingarian, 1976)

Table 3.5 *Forest area (10⁶ ha) and fuelwood, including charcoal, production (10⁶ t) in 1987*

	Total forest area[a]	Productive forest area[b]	Fuelwood (including charcoal production)
AFRICA			
Algeria	2.2	1.2	1260
Angola	54.0	21.0	2938
Benin	3.9	1.1	3173
Botswana	33.0	0.2	851
Burundi	–	–	2715
Cameroon	26.0	20.0	6894
Cen. African Rep.	36.0	19.0	2181
Congo	21.0	14.0	1202
Côte d'Ivoire	9.9	5.1	6098
Egypt	–	–	1433
Equatorial Guinea	1.3	1.0	319
Ethiopia	27.0	4.5	27243
Gabon	21.0	20.0	1874
Ghana	6.6	3.0	6271
Guinea	11.0	4.3	2735
Kenya	2.5	1.2	23959
Liberia	2.0	1.9	3346
Libya	0.3	0.1	382
Madagascar	14.0	7.2	4600
Malawi	4.3	0.6	4731
Mali	7.2	1.3	3477
Mauritania	–	–	5
Mauritius	–	–	15
Morocco	3.6	2.3	935
Mozambique	16.0	3.8	10189
Namibia	18.0	2.0	–
Niger	2.6	0.3	2789
Nigeria	15.0	4.5	67057
Réunion	–	–	22
Rwanda	0.3	0.1	4000
Senegal	11.0	1.8	2585
Sierra Leone	2.1	0.4	5663
Somalia	9.1	0.1	3253

Table 3.5 (continued)

	Total forest area[a]	Productive forest area[b]	Fuelwood (including charcoal production)
South Africa	1.3	1.1	5054
Sudan	48.0	32.0	13374
Swaziland	0.2	0.1	400
Tanzania	42.0	11.0	16577
Togo	1.7	0.4	457
Tunisia	0.4	0.4	1990
Uganda	6.1	2.0	8311
Zaire	178.0	139.0	21222
Zambia	30.0	6.6	6724
Zimbabwe	20.0	0.7	4383
Total Africa	**688.6**	**335.3**	**282687**
AMERICA, LATIN			
Argentina	45.0	34.0	4109
Belize	1.4	1.0	90
Bolivia	67.0	37.0	878
Brazil	518.0	423.0	125121
Chile	8.4	7.5	4470
Colombia	52.0	40.0	10823
Costa Rica	1.8	1.2	1914
Dominican Rep.	0.6	0.6	697
Ecuador	15.0	11.0	4501
El Salvador	–	–	3550
French Guiana	7.8	6.7	47
Guadeloupe	–	–	11
Guatemala	4.6	3.0	5048
Guyana	19.0	14.0	13
Haiti	–	–	4261
Honduras	4.0	3.0	3448
Jamaica	–	–	9
Martinique	–	–	7
Mexico	49.0	27.0	10369
Nicaragua	4.5	4.1	2063
Panama	4.2	2.9	1219
Paraguay	20.0	8.8	3632

Table 3.5 (continued)

	Total forest area[a]	Productive forest area[b]	Fuelwood (including charcoal production)
Peru	71.0	44.0	4659
Puerto Rico	0.3	0.1	–
Surinam	15.0	13.0	10
Trinidad & Tobago	0.2	0.2	16
Uruguay	0.6	0.2	2165
Venezuela	34.0	19.0	501
Total America, Latin	**943.4**	**701.3**	**193631**
AMERICA, NORTH			
Canada	264.0	215.0	4712
USA	226.0	195.0	80984
Total America, North	**490.0**	**410.0**	**85696**
ASIA			
Afghanistan	1.2	0.6	3866
Bangladesh	1.1	1.0	19783
Bhutan	2.1	1.8	2103
Brunei	0.3	0.3	57
Burma	32.0	24.0	11756
Fiji	0.8	0.3	27
Hong Kong	–	–	133
India	59.0	45.0	164384
Indonesia	119.0	76.0	94146
Japan	24.0	23.0	419
Korea (South)	6.5	3.8	3138
Laos	14.0	6.0	2850
Malaysia	21.0	16.0	5648
Nepal	2.1	1.3	11366
Pakistan	2.6	1.4	14643
Papua New Guinea	38.0	14.0	3950
Philippines	9.8	7.2	21910
Sri Lanka	1.8	1.3	5810
Taiwan, China	–	–	46
Thailand	16.0	8.0	23539
Total Asia	**351.3**	**231.0**	**389574**

Table 3.5 (continued)

	Total forest area[a]	Productive forest area[b]	Fuelwood (including charcoal production)
AUSTRALASIA			
Australia	42.0	37.0	2056
New Caledonia	0.7	0.4	–
New Zealand	7.0	2.8	36
Total Austral-			
asia	**49.7**	**40.2**	**2092**
Centrally planned economies (CPEs) and former CPEs			
Albania	1.2	0.9	1148
Bulgaria	3.4	2.5	1263
China	128.0	119.0	12768
Cuba	1.6	1.1	1883
Czechoslovakia	4.4	3.7	1054
GDR	2.7	2.6	507
Hungary	1.6	1.2	2129
Kampuchea	13.0	6.7	3554
Korea (North)	4.8	3.0	2891
Mongolia	9.5	4.2	964
Poland	8.6	6.4	2678
Romania	6.2	5.9	3267
former USSR	740.0	535.0	61975
Vietnam	10.0	5.4	16083
Yugoslavia	9.1	8.5	3193
Total CPEs	**944.1**	**706.1**	**115357**
EUROPE, WESTERN			
Austria	3.7	3.2	1009
Belgium	0.7	0.60	384
Cyprus	–	–	16
Denmark	0.5	0.40	290
Finland	20.0	18.00	2149
France	14.0	13.00	7447
FRG	7.0	6.80	2610
Greece	2.5	1.80	1438
Ireland	0.4	0.30	33
Italy	6.3	3.70	3217

Table 3.5 (continued)

	Total forest area[a]	Productive forest area[b]	Fuelwood (including charcoal production)
Netherlands	0.3	01.20	83
Norway	7.6	6.60	656
Portugal	2.6	2.20	427
Spain	6.9	6.50	2420
Sweden	24.0	22.00	3159
Switzerland	0.9	0.70	631
Turkey	8.9	6.60	7497
United Kingdom	2.0	2.00	105
Total, Europe, Western	**108.3**	**95.60**	**33571**
MIDDLE EAST			
Iran	3.8	1.61	708
Iraq	1.2	0.90	66
Israel	–	–	8
Lebanon	–	–	334
Syria	0.2	0.10	11
Yemen	–	–	218
Total Middle East	**5.2**	**2.61**	**1345**
WORLD TOTALS	**3581.6**	**2522.11**	**1103953**

[a] Total forest area applies to closed forest for temperate regions and to natural forest and plantations for tropical regions
[b] Productive forest includes total forest area which is accessible and able to sustain wood production. It excludes areas reserved in national parks and nature reserves

(Source: WEC, 1989)

Biomass other than wood

Biomass other than wood is seldom a commercial fuel and is rarely, if ever, traded on the international market. For this reason, statistics on potential fuels such as agricultural waste dung cakes, coconut shells and straw are not compiled or kept up-to-date. The amount of work involved in collecting information of this kind means that much of the published data reflect sample household surveys extrapolated to provide national averages.

Table 3.6 contains a review of the estimated quantities of dung and crop residues, per capita, in existence in 1979, projected to give an estimate for 1990. Table 3.6 indicates that in some countries these fuels are more readily available than sustainable forest yield.

Municipal, industrial and mining wastes

Reliable and comparable estimates of quantities of municipal waste are not available. Statistics on waste quantities are often collected, from city to city or region to region, under different categories and headings. The production of this waste is most likely related to population size, quantity of manufactured goods produced and local policies or necessity for re-use and recycling. Table 3.7 has been compiled using information from WEC (1989) listed under a number of headings including MSW, urban waste, municipal household waste, urban solid waste, solid household waste and, simply, waste, indicating a lack of international agreement on a term for domestic and institutional waste (US Department of Energy, 1988).

No figures are available for existing quantities of industrial wastes, but the sources and types of industrial wastes in the United States are presented in Table 3.8. It has been estimated that industrial waste makes up to 90 per cent of Russia's total waste stream (Pierce, 1993). It appears that, where these wastes are used for energy purposes, it is by the industry which produced them. Information on amounts of mining waste was not found, although it has been estimated that up to 10–15 per cent of mined materials may be lost in the form of fines or dust during mining, transport and processing. This represents considerable potential for energy recovery (see Chapter 6 for a discussion of briquetting).

Table 3.6 *Estimated organic resources, 1990 (GJ/capita/yr)*

	Sustainable forest yield	Dung[a]	Crop residues	Total
Algeria	3	3 SC	1	7
Egypt	0	1 CH	6	7
Iraq	1	5 CS	2	8
Bangladesh	2	4 CS	4	10
Ethiopia	3	6 SH	3	12
Kenya	1	7 CS	4	12
Ghana	8	2 CS	2	12
Nigeria	8	3 CS	2	13
Pakistan	1	6 CS	7	14
Morocco	6	5 CS	5	16
India	6	5 CS	6	17
Sri Lanka	11	2 CS	4	17
Korea, Republic	9	1 CP	7	17
Vietnam	12	1 CP	6	19
Afghanistan	6	8 CS	7	21
Philippines	12	3 CP	7	22
China	11	3 CP	8	22
South Africa	2	8 CS	13	23
Nepal	15	11 CS	6	32
Thailand	22	3 CP	9	34
Iran	22	6 SC	7	35
Korea, Democratic People's Republic	24	1 CP	13	38
Mexico	39	9 CH	9	57
Tanzania	54	11 CS	3	68
Indonesia	63	1 CS	6	70
Chile	71	7 CS	4	82
Burma	82	4 CP	6	92
Malaysia	114	2 CP	7	123
Sudan	148	18 CS	5	171
Argentina	104	45 CS	33	182
Colombia	180	16 CH	4	200
Venezuela	211	11 CH	4	226
Brazil	229	15 CP	9	253
Peru	245	7 CS	2	254

[a] Primary sources indicated by letter codes: C = cattle, buffalo, camel;
  H = horses, mules, asses; P = pigs; S = sheep, goats

(Source: IDRC, 1986)

Table 3.7 *Estimates of quantities of municipal waste used for energy production and direct energy produced from municipal waste*

Country	Quantity of waste per annum million tonnes	Total amount of direct energy produced per annum terajoules
Canada	unknown	8770.0
Chile	unknown	628.0
Finland	3.1	unknown
Israel	3.0	0.8
Italy	11.5	200.0
Norway	1.0	180.0
Portugal	2.0	unknown
Sweden	30.0	unknown
former USSR	50.0	2000.0
United Kingdom	25.0	1900.0
USA	unknown	50640.0

(Source: Compiled from SR-F, 1990; WEC, 1989)

Table 3.8 *Sources of industrial waste*

Industry	Description
Chemical	Contaminated wastewaters, spent solvent residuals, still bottoms, spent catalyst, treatment sludges, filter cakes
Fabricated metals	Electroplating wastes, sludges contaminated with metals and cyanides, degreasing solvents
Electrical equipment	Degreasing solvents
Petroleum refinery	Leaded tank bottoms, slop oil emulsion solids
Primary metals	Pickle liquor, sludge with metal contaminants
Transportation equipment	Degreasing solvents, metal sludges
National security	All types of wastes

(Source: EPA, 1986)

References

BELOSELSKY, B S (1991) *Personal Communication*. Moscow Institute of Power Engineering, Moscow.

BORD NA MÓNA (1985) *Fuel Peat in Developing Countries*. World Bank, Washington.

CHRISTOV, C and TODORIEV, H (1989) *Low-grade Coal Resources, Production and Utilization: Report 1989*. World Energy Conference, London.

COUCH, G R (1989) *World Survey of Low-grade Coal Resources, Production and Characteristics*. World Energy Conference, London.

COUCH, G R (1988) *Lignite Resources and Characteristics*. IEA Coal Research, London.

GORE, A J P (1990) *Ecosystems of the World: 4A and 4B*. Institute of Terrestrial Ecology, Huntingdon.

INTERNATIONAL DEVELOPMENT RESEARCH CENTRE (1986) *Energy Research: Directions and Issues for Developing Countries*. IDRC, Ottawa.

PIERCE, N (1993) 'Waste management challenges in Russia, Ukraine and Estonia'. *Waste Age* **24**: 194–198.

SMITH, A J (1981) 'Geological constraints on conventional energy resources'. In *Assessment of Energy Resources*. Report No. 9. Watt Committee on Energy, London.

SNAPE, C (ed) (1995) *Composition, Geochemistry and Conversion of Oil Shales*. Kluwer Academic Publishers, Dordrecht.

SPEDDING, P J (1988) 'Peat'. *Fuel* **67**: 883–890.

SR-F (199O) *Solid Waste Management in Sweden*. Svenska Renhållningsverks-Föreningen, Malmö.

US DEPARTMENT OF ENERGY (1988) *Five Year Research Plan: Biofuels and Municipal Waste Technology Program*. Department of Energy, Springfield.

US ENVIRONMENTAL PROTECTION AGENCY (1986) *National Screening Survey of Hazardous Waste Treatment, Storage, Disposal and Recycling Facilities*. US Environmental Protection Agency, Washington.

WORLD ENERGY CONFERENCE (1989) *1989 Survey of Energy Resources*. World Energy Conference, London.

YEN, T F and CHILINGARIAN, G V (1976) *Oil Shale*. Elsevier, Amsterdam.

4

Environmental management during the mining of low-grade fuels

A review of the approaches to gaining, harvesting or mining low-grade fuels is presented below along with techniques or practices for reducing negative environmental impacts. The low-grade fuels covered here range from minerals buried in the earth to organic substances left behind following crop harvest. For this reason it is difficult to make broad statements concerning the mining of low-grade fuels and the resultant effect on the existing environment.

In general, however, the mining of these fuels often has a greater impact per unit of energy produced than the mining of fuels such as anthracite or petroleum due to the lower energy density of low-grade fuels. Large tracts of land will be disturbed when practising opencast mining of lignite or oil shale, and underground mining of these fuels will be more extensive. There will be a greater number of workers required to mine low-grade fuels and, while this will create more job opportunities, it will also increase the likelihood of injury, disease and death due to mining activities. Land requirements for the build-up of solid waste and its disposal will also be greater when mining low-grade fuels.

New developments such as mines require appropriate infrastructure which may result in the building of roads and settlements in previously uninhabited areas or may bring large numbers of workers and their families to the region. The effects of this disruption of the existing community and land use must be taken into account when assessing the impacts of a mining project (Guthrie-Jones, 1991; British Coal Opencast, 1990; Coppin, 1989; Muskett, 1987; Chadwick, Highton and Lindman, 1987; Chadwick, Highton and Palmer, 1987; Selman, 1986). As with any development, it is important to identify the major impacts associated with an undertaking and to develop a detailed prevention and mitigation plan to reduce the possibility of environmental degradation.

Mining of lignite

The majority of lignite deposits are found in relatively shallow sedimentary basins formed during the last 200 million years (see Chapter 3, Figure 3.1) and are accessible through opencast mining. The greatest areas of concern during the mining of lignite include the loss of groundcover and existing land use, the health and safety of the mine workers, displacement and/or disruption of residents in the vicinity of the site and a potential decline in water quality and soil fertility.

Destruction of ground cover

With any opencast mining the ground cover must be removed, which leads to the loss, at least temporarily, of the existing land use. This impact may be relatively minor, for example, where the ground cover consists of derelict land, or it may have a more severe effect, for instance, the destruction of the habitat of a rare plant species or the removal of a family home. The major impacts accompanying a loss of ground cover are outlined below (Table 4.1) along with proposed mitigation or prevention measures. An example of a mining practice which will reduce the negative effects of opencast mining is presented in Figure 4.1.

Worker health and safety

Opencast mining of lignite can result in injury or death to mine workers.The formation and enforcement of a strict safety code at the site will significantly

Table 4.1 *Potential environmental impacts from opencast mining of lignite and oil shale*

Potential impact	Environmental protection measure
▪loss of existing land use (e.g. agriculture, forestry, recreation)	▪development of detailed plan for post-mining restoration of area
▪loss or disruption of wild-life habitat	▪design site to avoid important wildlife habitat; relocate wildlife to a similar environment; expand area of habitat in vicinity of site
▪loss or disruption of rare or endangered plant species	▪conduct survey of flora prior to site development; avoid area of rare or endangered plant species; relocate rare or endangered plant species to an appropriate site
▪loss of topsoil and sub-soil	▪store topsoil and subsoil in an area adjacent to the site; cover stored soil during early stages and seed with fast-growing plant cover to prevent erosion; reapply stored topsoil and subsoil as soon as possible following mining

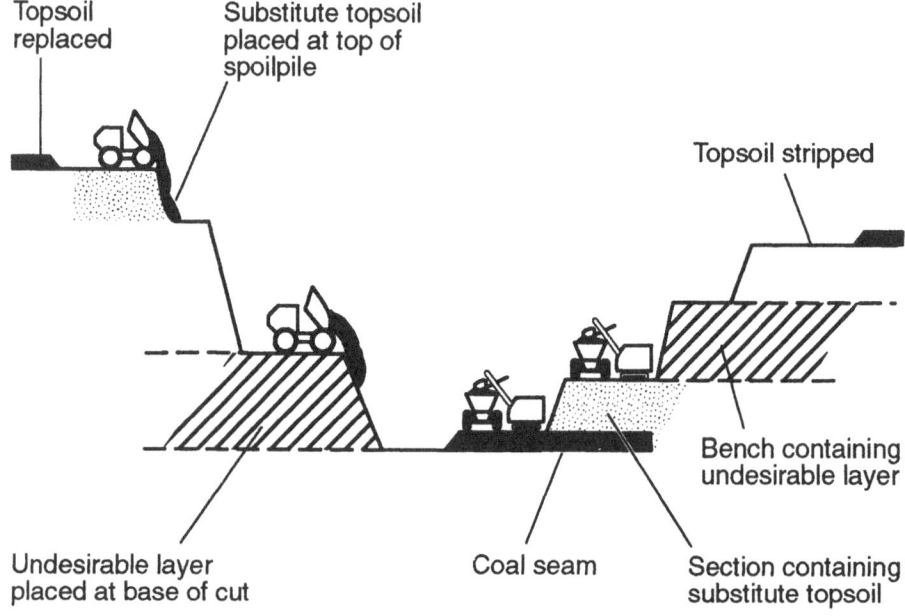

Figure 4.1 *Selective placement of overburden in truck and shovel operations of an open pit mine*

(Source: Hannan and Bell, 1986)

reduce the chance of mining-related accidents and hazards. The probability of injury or death during lignite mining is dependent upon a number of factors including:

- site size;
- size location;
- type of machinery used;
- quantity of material mined;
- number of workers per unit of material produced.

As fatalities in coal mining are generally ten times more likely in underground mines than surface mines, lignite mining, usually by surface methods, is less hazardous than underground mining for bituminous coal or anthracite.

Dust at the mining site may pose health concerns particularly with respect to respiratory ailments. This is especially true if the dust has a high silica content. Risk can be lessened, however, through the use of high-quality filter masks and reduced hours of exposure during periods of high wind. The use of water sprays on driveways and surrounding areas decreases this hazard. The health risk of the coal cycle is presented in Table 4.2.

Table 4.2 *Health risk of coal cycle: mining (data in fatalities/gigawatts (electric) (GW(e))*

Source	Nature of risk	Occupational	Public
Underground	immediate	1.2–1.5[a,c], 1.67[b,d], 1.24[e]	
	delayed	0.61–1.46[c], 0.24[d]	
Surface	immediate	0.15–0.17[c], 0.123[d], 0.09[e]	

Occupational
- Fatal accidents in coal mines, mainly from roof cave-ins and machinery
- Injuries to miners, mostly from material handling
- Miners' diseases: exposure to dust, leading to pneumoconiosis (underground mining), silicoses, bronchitis, vasoneurosis, vibration disease associated with use of equipment, leading to muscle and bone pains, bursitis and Raynaud's syndrome (white finger disease)

Public
- The principal health risk (mainly delayed) is from exposure to emissions in the conversion phase
- Disposal of coal mining and processing waste under some circumstances poses public risk of landslides or floods
- Not calculated

[a] From recent accidental fatility rates in North American and European mines
[b] The coal is assumed to contain 9% ash and 2% sulphur (UNEP, 1985b); the information on the conversion stage is based on a 1000-MW(e) power plant with 97% ash removal and no FGD
[c] Hamilton, 1984
[d] UNEP, 1985
[e] Hubert et al., 1981

(Source: IAEA, 1991)

Prolonged exposure to loud mining equipment can have a negative effect on worker hearing. In Germany, where up to 90 decibels (dB(A)) is considered a safe noise level, 39 per cent of workers were found to be mining in areas where the noise level fell between 91 and 101 dB(A) (UNEP, 1985a). The use of industrial strength ear covers and increasing the distance between noisy operating equipment and miners may prevent hearing loss. These impacts and environmental protection measures are summarized in Table 4.3.

Social impacts of lignite mining

Displacement of residents on mine site The greatest social impact from opencast mining may be the removal of residents from areas where mine

Table 4.3 *Health and safety during lignite mining*

Potential impact	Environmental protection measure
Worker health and safety	
▪risk of injury, disease and death	▪ensure high safety standards at the site, including protective clothing, footwear(e.g. steel-toed boots), hard hats and a stringent safety code regarding proximity of workers to moving machinery and blasting area; reasonable length of work shift with regular breaks to reduce fatigue-related accidents; industrial strength filter masks to reduce intake of dust containing potential respiratory irritants; industrial earmuffs to block out noise

development is to take place. This can cause distress to individuals and families due to the pressures of a forced move and feelings of a loss of control over their own decision making. Long-time area residents may be particularly affected both by the loss of their homes and, at least temporarily, a loss of a sense of community. Conversely, the remaining community may suffer from the loss of active and long-time community members.

Compensation packages offered to those displaced due to mining activities should include, aside from a financial settlement, services such as aid in locating a new home and organizing a move as well as counselling for those who feel it would be useful.

Disruption of residents in the vicinity of the mine   Residents in the vicinity of the site may be disturbed during activities of daily living by noise from blasting and on-site operations (Walker, 1989) and from increased levels of dust and traffic. They may also experience effects from workers moving into the area to provide labour for mining. To a certain extent the severity of these impacts is related to the scale of mining operations. A larger mine requires a larger workforce and will be operating in the community for a longer period of time.

It is particularly important to consider the social impact of an influx of workers into an area of low population density, where the laying of new roadways and the build-up of infrastructure needed to supply workers and their families may affect the existing character of the community. Changes in community character may not always be for the worse, but a decision in this regard is best made following discussions with community leaders and public information meetings where members of the community-at-large are both informed about the project and invited to give their views.

Methods for mitigating negative social environmental effects from mining (presented in Table 4.4) can include setting up a public liaison committee that comprises community leaders (representing a range of community interests) and officials from the mining company. The committee may be useful in maintaining an open and informed dialogue between representatives from the mine and local residents, resolving disputes and complaints concerning mine operations and, if appropriate, in developing a reasonable compensation package for residents adjacent to the site and the community as a whole.

Table 4.4 *Social impacts and mining of lignite*

Potential impact	Environmental protection measure
Displacement of residents	
■distress and loss of a sense of community	■offer an appropriate compensation package including a financial settlement and services to ease transition of move
Disruption to residents in vicinity of opencast mining	
■increased levels of dust	■installation of tall sprayers at site boundary; cease mining operations during times of high wind; wet all site roads with sprayer trucks at regular intervals; spray wheels of large dump trucks prior to site departure
■increased levels of noise	■operate during regular daylight hours; use low-impact explosives when possible; determine transportation routes to and from site with least impact on residents in the area
■vibration	■conduct structural survey of buildings adjacent to site; conduct blasting experiments prior to the initiation of mining activities to determine appropriate levels of underground vibration; monitor vibration from blasting at all times to ensure maximum no greater than 12–15 mm/sec at closest structures

Table 4.4 (continued)

Potential impact	Environmental protection measure
■increased traffic	■assess travel routes to and from site and develop plan to avoid built-up areas, schools and other facilities likely to be disrupted by increased traffic from the mine; travel during times of low traffic levels to decrease risk of accidents
■influx of mine-related labour force	■determine and plan for increased demand on community goods and services such as education and housing of labour force

Water quality and soil fertility

Water escaping from the mining site may have a negative effect on nearby bodies of surface water. Acid mine drainage can alter the properties of soil through which it passes and pollute streams in the vicinity of the mine, leading to loss of biological organisms and communities.

Ensuring containment and treatment of water from the mining operation is the best approach to eliminating these types of impacts. Low-cost variations include using a fountain to spray water into the air where oxidation of iron and suspended solids will occur or building a series of sedimentation basins (holding ponds) at different elevations, where water from the site will cascade from one pond to the next allowing oxidation (Parker, 1992; Glover, 1975). Holding ponds to which neutralizing agents such as limestone have been added may be used where pollution levels are high. Also, for high acidity or where it is not possible to construct holding basins, technology such as an in-line aeration system (Figure 4.2) may be installed at the site.

An additional problem of water leaving the mining site is the erosion associated with this run-off and the resultant increased sediment transport in nearby streams. The additional sediment load can also reduce the range of plant and animal species able to thrive in adjacent water bodies and, under conditions of high water, may contribute to flooding. This problem may be controlled through the use of retaining walls and closed systems of water recirculation. These issues are summarized in Table 4.5.

Interaction with groundwater

Any mine pit which is dug below the level of groundwater in the area will cause water to drain out of the water-bearing levels. We have noted above

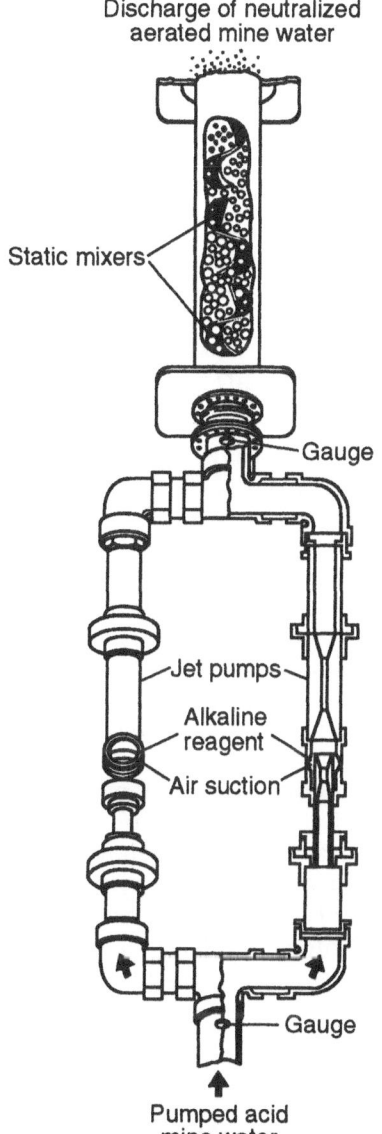

Discharge of neutralized
aerated mine water

Static mixers

Gauge

Jet pumps

Alkaline
reagent

Air suction

Gauge

Pumped acid
mine water

Figure 4.2 *In-line aeration and neutralization system*

(Source: Carter, 1989)

Table 4.5 *Water and soil quality in areas of lignite mining*

Potential impact	Environmental protection measure
Decline in water quality and soil fertility	
■addition of sediments to streams	■prevent run-off from site through the use of holding ponds and walls or closed system water recirculation
■acid mine drainage	■monitor pH levels of water draining from site to ensure a pH level within 5–9; a lower pH level indicates water should be controlled and treated by the installation of an in-line aeration and neutralization system (Figure 4.2), through treatment with alkaline chemicals such as crushed limestone to neutralize acid water or by filtering through adjoining wetlands that have been tested and found to improve water quality without a negative impact on the existing plant communities

that the subsequent disposal of this water can have deleterious impacts upon the local environment. In certain cases, the drainage of water can also lower the water table in the surrounding area, with potentially harmful results.

The scale of this impact will depend upon the volume of water drained, the size of the original aquifer and the use which is made of the water. In general, only very large mines will usually cause significant drops in the water table (Noetstaller, 1987). However, in certain cases the impacts may be major. Preliminary hydrological surveys and calculations need to be made in advance of the mine's development to estimate the scale of the problem.

Site restoration

Usually, lignite is mined by surface-mining methods, specifically, strip- or opencast mining. Generally, the mining procedure begins by pumping out the groundwater from the area to be mined. Following this, a well-defined sequence of events should be initiated so that after the lignite has been removed, or sequentially during its removal, the topsoil, subsoil and overburden material (usually rock or partially weathered material) can be replaced in the reverse sequence. Thus an opening, pioneer cut is made in the surface, and topsoil is removed for storage; overburden is removed and stored separately. Following the removal of the lignite the sequence is repeated, but the overburden this time is placed in the mined-out area (Figure 4.3). This process is necessary as the overburden may represent in excess of four times the volume of lignite gained.

Generally, the overburden material, or spoil, associated with lignite is, or on exposure quickly becomes, acid (pH 4.5–2.0). It is low in available plant nutrients and is also water-repellent. This lack of wettability means that high run-off and erosion can occur from the stockpiled heaps of spoil.

Before a productive plant cover can be established, there must be replacement and grading of the overburden in the worked-out area of the strip mine and, at the same time, the acidity, low fertility and wettability problems need to be corrected. There are now well-established methods of dealing with all of these problems (Knabe, 1973, 1959; Schulze and Engels, 1962). Lime or limestone (CaO or $CaCO_3$) is applied as the spoil, topsoil and subsoil are replaced, being worked in by rotovation. Up to 200 tonnes per hectare may be necessary. Use of ash from the boilers where lignite has been burnt will allow the wettability to be improved. It must be worked into the spoil material. Large applications of NPK + Mg fertilizer are also required – up to 500 kilograms per hectare. This form of treatment should allow tree and crop yields equivalent to those obtained before mining, but where little previous experience is available to suggest the most suitable native species to use, simple species trials will need to be undertaken.

As with all mining land restoration, clear land use objectives should be identified, long in advance of the restoration work. The operation should have a clearly specified sequence of events; the spoil and soil material should be analysed, and the amelioration and planting specification planned. Restored sites need to be closely monitored, and maintenance work (extra fertilizer, repeat seeding or planting, extra drainage) may have to be undertaken in the first few years following the restoration. Sequences required to bring about a successful reclamation scheme are given in Figure 4.4. They are relevant to land restoration following the mining of other fuels as well.

Oil shale mining

It is possible to mine oil shale through opencast (strip-mining) methods and, where this is the case, approximately 80–90 per cent of the oil shale will be recovered. The use of underground mining techniques will result in a retrieval of 35–50 per cent of the deposit (UNEP, 1985b).

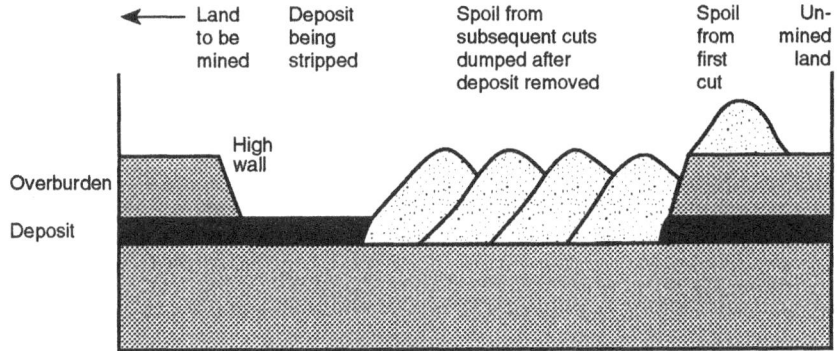

Figure 4.3 *Strip mining for lignite*

(Source: Bradshaw and Chadwick, 1980)

Impacts from the surface mining of oil shale have some similarities to those identified for lignite opencast mining (see Tables 4.1, 4.3, 4.4 and 4.5) but in many instances the scale or intensity of these impacts will be greater. Larger tracts of land will be disturbed for mining purposes due to the high quantities of oil shale needed from an operation, and blasting may be required to fracture the oil shale into pieces suitable for combustion or retorting.

The surface disruption associated with underground mining of oil shale would obviously be less than the effects of opencast mining, but there would still be blasting and traffic concerns for residents in the vicinity of the site, and sinking of ground (subsidence) may also occur above areas of underground mining. As mentioned earlier, the overall efficiency of recovery is also markedly less with underground oil shale mining.

Acidic run-off

The problem of acid drainage from the mining site can be acute with oil shale mining, depending on the components of the oil shale deposit. Oil shales with a high pyrite (FeS_2) content are particularly hazardous to surface water ecosystems located near the mine, since oxidized FeS_2 may reach adjacent water bodies in the form of ferric sulphate. Where this occurs the ferric sulphate can combine with water to become sulphuric acid, which may severely reduce pH levels and, depending upon the concentration, release trace elements present in the shale or stream sediments.

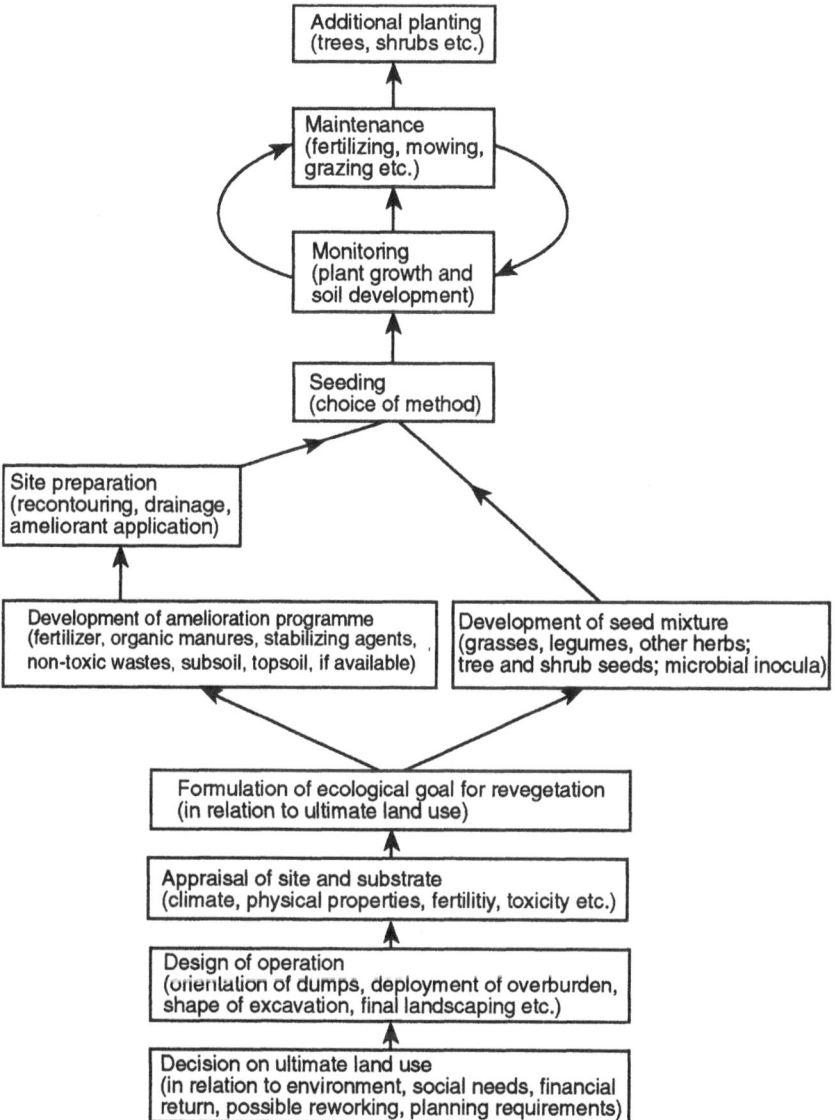

Figure 4.4 *Sequence of actions and decisions necessary for planning successful reclamation during or following mining*

Source: Bradshaw and Chadwick, 1980)

Land restoration

Two separate aspects of land disturbance accompany the mining and conversion of oil shale. Where strip mining is used to obtain the oil shale material, this land requires restoration to biological productivity. But in addition, much land is required to dispose of the spent oil shale and ash (obtained by roasting at a higher temperature) after it is roasted. In this process the oil shale undergoes changes that result in a spent shale that is greater in volume than the original material.

The strip-mined land can be reclaimed by planting into the graded overburden material, either uncovered or covered with soil stored during the mining process. Successful reclamation schemes in Estonia have demonstrated the feasibility of this.

The spent oil shale and ash are quite different in the chemical and physical characteristics of the substrate than the mined overburden. Spent shale and shale ash have values between about pH 7.0 and 12.0 due to high concentrations of soluble salts. Solutions to the problems of reclaiming spent oil shale are listed in Table 4.6.

Table 4.6 *Suggested solutions to enable plant establishment and growth problems associated with spent oil shale reclamation to be overcome*

Problem	Suggested action
∎Material difficult to wet	∎Mix spent shale with water mechanically at time of disposal
∎High pH and salinity	∎Remove salts by slow leaching
∎Ion imbalance	∎Apply gypsum ($CaSO_4$)
∎Low fertility	∎Apply nitrogen and phosphorus fertilizer

Ferric sulphate may also form a precipitate of ferric hydroxide (termed 'yellow boy') which can form a photosynthesis-inhibiting layer in the stream (Hutchinson, 1957). Measures for reducing and preventing these impacts of oil shale mining are summarized in Table 4.1.

Worker health and safety during oil shale mining

Opencast mining is generally safer for oil shale miners than underground extraction. The risk of fatal injury from underground mining of oil shale is roughly the same as coal, which is 0.5 deaths for each million tonnes of mining material where the average annual output per worker is 3000 tonnes (Pochin, 1976). This same output, under conditions of opencast mining, has a risk factor of 0.1 deaths.

Factors influencing the potential for safe working conditions during opencast mining of oil shale are the same for lignite and are described in this chapter, as are concerns regarding dust and noise levels during mining. There is some evidence suggesting an increased potential for skin and, possibly, lung cancer in oil shale workers, but this relates more strongly to the carcinogens forming during oil shale processing than during mining. A summary of the health risks associated with the oil shale fuel cycle is given in Table 4.7.

Table 4.7 *Health risk of oil shale cycle: mining (data in fatalities/GW(e))*

Source	Nature of risk	Occupational	Public
Underground	immediate	0.28[a,b], 244[c,d]	
Surface	immediate	40.7[c,d]	

Occupational
  – Mining risks similar to coal

Public
  – Main risk is from exposure to air pollutants from burning waste heaps, retorting and combustion
  – Not calculated

[a] Hamilton, 1984   [c] UNEP, 1985b
[b] IWG Corporation, 1982   [d] House, 1981

(Source: IAEA, 1991)

Social impacts of oil shale mining

The social impacts of oil shale mining are of the same nature as those discussed for lignite in this chapter.

Harvesting of peat

Peat may be taken from its original site in either a wet or a dry form. Dry removal is known as harvesting and is the most common method of obtaining peat for fuel. Wet mining of peat is possible in very waterlogged peatlands, using hydraulic or mechanical methods, but this results in high moisture peats which require a good deal of dewatering to make them burnable.

Fuel peat is harvested by hand or machine in solid blocks known as sod peat. It may also be shredded or manufactured in a crumb or powdered form called milled peat. The production of milled peat is carried out in several stages, by machine, and makes use of solar drying. The peat is shredded in situ then turned repeatedly to hasten drying. Once the moisture level reaches 45 per cent, the peat is piled and stored in large mounds (up to 15 m high) or in long ridges (3–4 m high; 500–3000 m long) (Bord na Móna, 1985).

Harvesting of peat usually begins with an assessment of the deposit, determining the size and depth and the nature of the existing peat. This is accomplished by conducting a ground survey and taking samples. The characteristics of peat can vary throughout a deposit, and standard tests are often carried out to identify the range of a number of variables including moisture content, bulk density, ash content, calorific value and pH (Bord na Móna, 1985).

Impacts on the physical environment due to peat harvesting

Impacts due to the harvesting of peat include loss of ground cover and the associated loss of wildlife habitat and existing vegetation. The hydrological impacts of peat harvesting may pose the greatest disruption to the existing wetland ecosystem of which the peat deposit is a major component (Boron, Evans and Peterson, 1987).

Potential hydrological impacts include alteration or disruption of surface water flow patterns and changes in groundwater elevations. Effluent produced from the drainage of a wetland will often contain a high level of suspended solids which must be controlled to prevent flooding or pollution of nearby surface waters. The loss of the absorbency of peat may also increase the danger of flooding. These impacts along with appropriate mitigation measures are presented in Table 4.8.

As with lignite and oil shale, the scale of these impacts is related to the scale of harvesting operations. Table 4.9 estimates the magnitude of potential hydrology and hydrogeology impacts in New York State, USA, according to the quantity of peat harvested.

Table 4.8 *Impacts on the physical environment due to peat harvesting*

Potential impact	Environmental protection measure
Removal of ground cover	
■loss of existing land use	■development of detailed plan for post-harvesting land development
■loss or disruption of rare or endangered plant species	■conduct botanical survey to determine extent of rare or endangered plant species; plan site to avoid these areas where possible; attempt to transfer to similar habitat
■loss or disruption of wildlife habitat	■conduct survey to determine presence of wildlife habitat at harvesting site; avoid habitat areas where possible and keep site size to a minimum
Disruption of existing ground and surface waters	
■modification of ground-water elevation due to loss of water-holding peat	■construct holding basins to mimic the water detention functions of peatland
■increased floodwater flow	■use holding basins and gradual pumping of excess water from the site
■modification of nearby water flow patterns	■assess the impact of altered water flow patterns on local land use; if the impacts are undesired, use holding ponds to collect excess water
Changes in water quality	
■discharge of acidic water	■holding ponds where run-off treated to restore pH levels prior to release from site; buffer zone around site
■increased BOD/COD	■holding ponds where BOD/COD normalized through natural aeration and biological and chemical oxidation (Hazen and Beeson, 1979); buffer zone around site
■increased discharge of nutrients, colloidal and settleable solids, heavy metals	■holding ponds where wastewater will settle and unwanted substances can be removed and treated; buffer zone around site

Table 4.9 *Summary of potential hydrology and hydrogeology impacts from peat mining*

Impact	Regional		Site-specific	
	Small-scale[1]	Large-scale[2]	Small-scale	Large-scale
Increased floodwater flow potential	Minor	Moderate	Minor	Major
Groundwater elevation modification	Minor	Moderate	Minor	Moderate
Modification of surface flow patterns	Minor	Moderate	Minor	Moderate
Increased minimum stream flow rate	Minor	Minor	Minor	Minor
Increased mean stream flow rate	Minor	Minor	Minor	Minor
Altered hydrologic budget	Minor	Minor	Minor	Moderate
Altered ground water aquifer	Minor	Minor	Minor	Moderate
Reduced evapotranspiration	Minor	Minor	Minor	Minor

[1] Small scale = Production rate equals 100,000 dry tons per season, dry mining methods, mine life of 6–19 years
[2] Large scale = Production rate equals 300,000 dry tons per season, dry or wet mining methods, mine life of 5–6 years

(Source: Newman et al., 1985)

Worker health and safety during peat harvest

Cutting sod peat by hand for small-scale use, such as providing peat for space heating a single household, presents few concerns for worker health and safety, although care should be taken in using sharp spades and lifting heavy loads. Mechanical harvesting of peat is similar to agricultural harvesting using large machinery. The risk of injury (Table 4.10) is linked to knowledgeable and careful operation of machines with sharp blades and rapidly moving belts and mechanical parts for cutting, milling, harrowing, harvesting and loading the peat.

Table 4.10 *Health risks of peat cycle: harvesting (data in fatalities/ GW(e))*

Source	Nature of risk	Occupational	Public
Extraction	immediate	0.8–0.9[a,b,c]	

Occupational

−Potential respiratory and eye irritation from dust exposures during extraction and handling, which are high under certain weather conditions
−Spontaneous combustion during harvest

Public

−Respiratory diseases due to exposure to pollutants during combustion

[a] Includes bog preparation (0.1–0.2) and extraction (0.7)
[b] UNEP, 1985b
[c] Lautkaski et al , 1982

(Source: IAEA, 1991)

Per unit of energy produced, peat can require a greater number of workers than lignite and oil shale (Hamilton, 1992). This increases the possibility of injury or accident due to peat harvesting. Although, depending on prevailing weather conditions, there are considerable quantities of dust associated with peat production, disease due to dust exposure during peat extraction has not been found (Lautkaski et al., 1982). Since the operating machinery may be loud, industrial-strength ear covers are also required to prevent hearing loss. This review of worker health and safety issues relating to peat production is summarized in Table 4.11.

Table 4.11 *Health and safety during peat harvesting*

Potential impact	Environmental protection measure
Worker health and safety	
■risk of injury and death	■use of equipment by well-trained, safety-conscious operators only; strict guidelines regarding proximity of groundworkers to those in machines; use of industrial ear muffs for those operating loud machinery; training in safety and fire-fighting

Social impacts of peat harvesting

Residents in the vicinity of a peat-harvesting site may experience disruption of activities of daily living due to dust created during the cutting and milling of peat and noise from machinery operating at the site. This disturbance usually occurs during only one season of the year and is similar to the harvesting of agricultural crops in the disruption which is created. Dust may be controlled by ceasing harvesting operations when winds are high. The problems of noise may be reduced if operations only take place during daylight hours.

A change in the aesthetic quality of the landscape may occur due to the presence of peat stored in mounds up to 15 m high or ridges of peat hundreds of metres long.

Harvesting or collecting of wood

The harvesting of wood for fuel takes place on a variety of scales. Individuals may gather wood for use in a single household, or larger quantities of wood may be taken for commercial fuelwood production.

Negative environmental impacts associated with the collection or harvesting of wood for energy purposes are generally site specific and are dependent upon a number of factors including:

- location and size of adjacent human settlements;
- local soil and climate conditions; and
- cultural and economic importance of forested lands.

It has been estimated that over 2 billion people in developing countries use wood to cook their food (Goldemberg et al., 1988). Much of this wood consists of dead and fallen branches and twigs and is collected on an 'as needed' basis. Where trees are cut for fuel, a situation which is more likely to arise near larger urban settlements where fuelwood and charcoal may be sold as commercial products, a variety of environmental effects may occur (Foley, 1985).

The use of communal wooded areas managed to produce fuelwood is well established in certain areas of the developed world, although it is now carried out on a much smaller scale than in the past. These communal fuelwood lots can be found in a number of European countries and areas of North America settled by Europeans who practised the activity in their countries of origin. This sustainable form of specialized forestry is found in the Ardennes in Belgium and France, in Bavaria in Germany, in Austria, Nordic countries and elsewhere. Large stacks of wood associated with many dwellings are a common sight. The wood may originate from areas of the forest cleared on a rotational basis or, less frequently in the case of fuelwood, from woodland managed by coppicing.

The practice of coppicing, 'cutting and allowing regrowth', has revived the possibility of using land (particularly areas of wasteland) to provide wood (and other forms of biomass) as an energy plantation. To this end a number of surveys have been undertaken of the supply potential of such a strategy (Dennington, Chadwick and Chase, 1983; Steinbeck, 1980) and specialized harvesting machinery has been developed (Duff and McLain, 1980; Hakkila and Mäkelä, 1975).

Impacts to the physical environment due to harvesting of wood

Deforestation usually refers to the taking of wood from a forest at a rate which exceeds the ability of the forest to regenerate or replace by growth the amount harvested on an annual basis. Wide-scale deforestation, and the associated problems of desertification and erosion of topsoil, are rarely the outcome solely of cutting trees to obtain fuel. The most common cause of this type of deforestation is clearing of land for agricultural or building purposes (Kaale, 1990; Pasztor and Kristoferson, 1990; Goldemberg et al., 1988; Leach and Mearns, 1988; Eckholm, 1982).

The removal of trees is a disturbance of the groundcover which can lead to a decrease in the complexity of the existing plant community and destroy or disrupt wildlife habitat. Trees provide shade which can be particularly important in hot climates to give respite from the hot sun and to reduce evaporation of moisture from the soil. The root systems of the trees help to reduce the likelihood of soil erosion and aid in the distribution of nutrients deep into the soil. In some cultures, the aesthetic and recreational value of forests may be important, or the presence of wildlife or plants for food may be a necessary part of existence.

The management of forest resources through the development and enforcement of a reforestation programme is one approach to reducing the negative effects of using wood for fuel. In some areas, silviculture (the growing of trees) can be practised on land that is not suitable for food crop cultivation. These wood plantations usually consist of one or two rapidly growing tree species. Wood plantations are most successful where they represent a commercial interest near a densely populated area. In practice, wood from plantations is most frequently used for poles and building materials.

It is common to use species of *Eucalyptus* for woodfuel plantations due to its propensity for rapid growth. These species have been criticized for causing harmful impacts to the environment stemming from their heavy uptake of groundwater. In arid regions, this can create problems for nearby agriculture. It may also limit possibilities for the growth of inter-tree ground cover which can be used for pasture or forage.

A variety of strategies for dealing with local wood shortages has been proposed and implemented by a large number of international aid agencies, particularly over the past two decades. Foley (1991) has pointed out the inadequacy or failure of these projects to make a genuine difference in the manner in which fuelwood is consumed (e.g. widespread acceptance of cookstoves designed to conserve fuel) or forests maintained on a sustainable basis. This is seen as a result of a lack of understanding of the dynamics surrounding a change in the type of fuel used in the home or the method of employing the fuel. In many cases, rather than enter into a plan for local tree plantations or purchasing more fuel-efficient stoves which cost money and may even produce food which tastes differently than the traditional method, wood users will cook less and/or search for and use other available biomass.

Collection of biomass other than wood

The collection by hand of biomass other than wood for energy purposes has little direct effect on the environment, particularly if these materials are already seen as waste and are removed from agricultural areas as part of crop harvest practices (Eriksson and Prior, 1990; Barnard, 1985). Some woodlike biomass such as corn-cobs, millet stalks, jute sticks, coffee prunings and coconut shells and husks are slow to decompose (Kaale, 1990) and would generally be considered as refuse if they were not used for energy.

Certain types of agricultural residue do have alternative uses (e.g. straw for bedding animals) and it has been suggested that, when residues are burnt as they stand in fields, this provides valuable minerals to the soil which are not available when the residues are diverted to fuel use. There exists very little documented evidence for this, however.

In general, small-scale diversion of agro-residues is unlikely to cause any significant environmental problem. However, a large-scale diversion, for example, for briquetting, should be examined for potentially adverse impacts. Much will depend on the specific area from which residues are to be obtained.

The collection of large quantities of straw or stalks could raise dust which may affect human health (e.g. allergies, respiratory ailments) (Stjernquist, 1990). Poor storage of residues (e.g. not properly dried and baled) can also contribute to the development of mould, another potential health hazard (Jenkins and Summer, 1986).

Municipal, industrial and mining wastes

Municipal, industrial and mining wastes are not mined or harvested in the same manner as natural resource fuels. Initially, they may be considered for entry into the fuel cycle only after they have been identified as a disposal problem. In most industrialized countries, energy from waste is considered less important than recycling or landfill as a waste management strategy.

References

BARNARD, G W (1985) 'The use of agricultural residue as a fuel'. *Ambio* **4–5**: 259–266.

BORD NA MÓNA (1985) *Fuel Peat in Developing Countries*. World Bank, Washington.

BORON, D J, EVANS, E W and PETERSON, J M (1987) 'An overview of peat research, utilization and environmental considerations'. *International Journal of Coal Geology* **8**:1–31.

BRADSHAW, A D and CHADWICK, M J (1980) *Restoration of Land*. Blackwell Scientific Publications, Oxford.

BRITISH COAL OPENCAST (1990) *Opencast Coal Mining in Great Britain*. British Coal Opencast, Mansfield.

CARTER, R A (1989) 'Reclamation plans new ground'. *Coal* **26**: 44–48.

CHADWICK, M J, HIGHTON, N H and LINDMAN, N (1987) *Environmental Impacts of Coal Mining and Utilization.* Pergamon Press, Oxford.

CHADWICK, M J, HIGHTON, N H and PALMER, J P (1987) *Mining Projects in Developing Countries: A Manual.* Beijer Institute Centre for Resource Assessment and Management, York and Stockholm.

COPPIN, N (1989) 'Environmental assessment for opencast coal mining'. *Mine and Quarry Environment* **3**: 17–23.

DENNINGTON, V N, CHADWICK, M J and CHASE, D S (1983) 'Energy cropping on derelict and waste land'. *Journal of Environmental Management* **16**: 241–260.

DUFF, J W and McLAIN, H D (1980) 'Design of a harvester for 2–3 year old willow sticks'. In *Energy from Biomass* (ed by W Palz, P Chartier and D O Hall). Applied Science Publishers, London.

ECKHOLM, E (1982) 'Human wants and misused lands'. *Natural History* **91**: 33–48.

ERIKSSON, S and PRIOR, M (1990) *The Briquetting of Agricultural Wastes for Fuel.* FAO, Rome.

FOLEY, G (1991) *Energy Assistance Revisited: A Discussion Paper.* SEI, Stockholm.

FOLEY, G (1985) 'Wood fuel and conventional fuel demands in the developing world'. *Ambio* **4–5**: 253–258.

GLOVER, H G (1975) 'Acidic and ferruginous mine drainages'. In *The Ecology of Resource Degradation and Renewal* (ed by M J Chadwick and G T Goodman). Blackwell, Oxford.

GOLDEMBERG, J, JOHANSSON, T B, REDDY, A K N and WILLIAMS, R H (1988) *Energy for a Sustainable World.* John Wiley, New York.

GUTHRIE-JONES, D (1991) 'The Zhaotong study: exporting Australian know-how'. *Mining Magazine.* October.

HAKKILA, P and MÄKELÄ, M (1975) 'The Pallari bush harvester. *Finnish Forestry* 249.

HAMILTON, L D (1992) 'Health and environmental management of low-grade fuels'. In *The Environmentally Sound Management of Low-grade Fuels.* SEI, Stockholm.

HAMILTON, L D (1984) 'Health and environmental risks of energy systems'. *Risks and Benefits of Energy Systems.* IAEA, Vienna.

HANNAN, J C and BELL, L C (1986) 'Surface mine rehabilitation'. In *Australasian Coal Mining Practice* (ed by C H Martin). Australian Institute of Mining and Metallurgy, Parkville, Victoria, Australia.

HAZEN, C B and BEESON, B J (1979) 'Techniques for the assessment of bog hydrology, harvesting and mitigation of water quality impacts resulting from peat harvesting'. In *Proceedings of Management and Assessment of Peat as an Energy Resource.* Institute of Gas Technology, Arlington.

HOUSE, P W (1981) *Energy Technology and Environment.* US Department of Energy, Washington.

HUBERT, P, MOATTI, J P, MACCIA, C and FAGNANI, F (1981) *Les Impacts Sanitaires et Ecologiques de la Production d'Electricité: le cas français.* Centre d'étude sur l'évaluation de la protection dans la domaine nucléaire, Fontenay-aux-Roses.

HUTCHINSON, G E (1957) *A Treatise on Limnology.* Vol. 1. John Wiley, New York.

IAEA (1991) 'Comparative environmental and health effects of different energy systems for electricity generation'. In *Senior Expert Symposium on Electricity and the Environment: Key Issue Papers.* International Atomic Energy Agency, Vienna.

IWG Corporation (1982) *Health and Environmental Effects.* Documents on oil shale. IWG, San Diego.

JENKINS, B M and SUMNER, H R (1986) 'Harvesting and handling agricultural residues for energy'. *Transactions of the ASAE* **29**(3): 824–836.

KAALE, B K (1990) 'Traditional fuels'. In *Bioenergy and the Environment* (ed by J Pasztor and L A Kristoferson). Westview Press, Boulder.

KNABE, W (1973) 'Development and application of the Domsdorf ameliorative treatment on toxic spoil banks and lignite opencast mines, Germany'. In *Ecology and Reclamation of Disturbed Land* (ed by R J Hutnik and G Davis). Gordon and Breach, New York.

KNABE, W (1959) *Zur Wiederurbarmachung im Braunkohlenbergbau.* Deutscher Verlag der Wissenschaften, Berlin.

LAUTKASKI, R, POHJOLA, V, SAVOLAINEN, I and VUORI, S (1982) 'A comparative assessment of the health impacts of coal-fired, peat-fired and nuclear power plants'. *Health Impacts of Different Sources of Energy.* IAEA, Vienna.

LEACH, G and MEARNS, R (1988) *Beyond the Fuelwood Trap.* Earthscan, London.

MUSKETT, C J (1987) 'Environmental assessment of opencast coal development'. *Mine and Quarry Environment* **2**: 12–15.

NEWMAN, J R, DOOLITTLE, J D, BUFF, D A, HART, R, NEFF, C R, ZWOLAK, R A, REA, R H and SCHNEIDER, R L (1985) *Statewide Environmental Impact Evaluation of Peat Mining and Development.* New York State Energy Research and Development Authority, Albany.

NOETSTALLER, R (1987) *Small-scale Mining: A Review of the Issues.* World Bank, Washington.

PARKER, S (1992) *Personal Communication.* British Coal, South Yorkshire Group, Castleford.

PASZTOR, J and KRISTOFERSON, L A (1990) *Bioenergy and the Environment.* Westview Press, Boulder.

POCHIN, E (1976) *Estimated Population Exposure from Nuclear Power Production and Other Radiation Sources.* OECD, Paris.

SCHULZE, E and ENGELS, H (1962) *Rekultivierung von Lössboden im Rheinischen Braunkohlengebiet. Zeitschrift für Acker- und Pflanzenbau* **117**: 247–272.

SELMAN, P H (1986) 'Coal mining and agriculture: a study in environmental impact assessment'. *Journal of Environmental Management* **22**: 157–186.

STEINBECK, K (1980) 'Short rotation forestry as a biomass source: an overview'. In *Energy from Biomass* (ed by W Palz, P Chartier and D O Hall). Applied Science Publishers, London.

STJERNQUIST, I (1990) 'Modern wood fuels'. In *Bioenergy and the Environment*. Westview Press, Boulder.

UNEP (1985a) *The Environmental Impacts of Production and Use of Energy*. Part 4. United Nations Environment Programme, Nairobi.

UNEP (1985b) *The Environmental Impact of Exploitation of Oil Shales and Tar Sands*. UNEP, Nairobi.

WALKER, A (1989) 'Opencast mining noise'. *Mine and Quarry Environment* **3**: 20–22.

5

Environmental management of the transport of low-grade fuels

The transport of low-grade fuels can take many forms and may occur more than once during the fuel cycle (e.g. from mine to processing plant, from processing plant to place of conversion). Lignite or oil shale may be moved from the mine mouth to a nearby power plant by rail or conveyor belt. These fuels, as well as peat, industrial, municipal or mining waste, may be transported by truck, rail or barge.

Environmental impacts and low-grade fuel transport

The environmental impacts associated with the transport of wood, wood other than biomass and, in many cases, peat for fuel purposes are generally minimal, because these fuels are frequently transported on a small scale for household use by hand, animal power or a small-engined machine. Low-grade fuels are also transported less frequently and over shorter distances than other fuels because of their high content of inert material. Under most circumstances it is considered too costly to transport large quantities of lignite or oil shale long distances when such a great percentage of the material cannot be used to produce energy (Couch, 1990b; Chadwick, Highton and Palmer, 1987). Most coals, including lignite, are utilized near the mine unless specific characteristics make long-distance transport attractive. Powder River Basin sub-bituminous coals in the western United States can compete with bituminous coal due to their low sulphur content and low mining costs. These factors make their use economically viable as far as 3000 km from source (Huettenhain, 1992).

Wood and sod peat give off very little dust in transport and therefore have a relatively lower impact on air quality than lignite, oil shale, milled peat and mining waste. Municipal and industrial waste can escape from freight cars, trucks or conveyor belts during transit, causing an eyesore and a build-up of those wastes in surface waters and across the landscape. Major impacts resulting from low-grade fuel transport are most likely to occur during large-scale operations such as supplying lignite to a plant which produces energy all year round. For this reason, the discussion in this chapter concentrates on large-scale transportation of low-grade fuel.

Little research has been undertaken specifically to identify the risk to the environment posed by the transport of low-grade fuels. This may be because the impacts of small-scale transport of low-grade fuels are generally minimal, and large-scale applications are not widespread. The use of peat and lignite to produce electricity is growing, and it seems reasonable to assume that moving large quantities of peat, lignite and oil shale would be similar to the transport of higher-grade solid fuels such as coal, about which more is known (Hamilton, 1992; Hamilton, 1979; Morris, Novak and Hamilton, 1979; Szabo, 1978).

Storage of low-grade fuels

Transport is a temporary form of storage which is an important consideration, since lignite and peat are known to be self-heating during containment. This fact, combined with their high reactivity, can lead to spontaneous combustion or explosion (Lishtvan, Gavrilchik and Falyushin, 1985; Laine, 1983). Generally, the lower the moisture content and the smaller the particle size, the greater the likelihood of fire-related accidents (Mutanen, Nissinen and Linna, 1990). Table 5.1 provides an overview of factors affecting the ignition and explosion of peat, wood, coal and lignite dust.

Table 5.1 *Typical explosion and ignition properties of peat, wood, coal and lignite dust (at atmospheric pressure)*

Property	Peat	Wood	Coal	Lignite
Maximum explosion pressure (bar)	8.3	9.2	7.6	8.0
k_{st}-value (bar m/s)	141.0	142.0	98.0	121.0
Minimum explosible concentration (g/m³)	45.0	30.0	55.0	30.0
Limiting oxygen concentration (%, nitrogen (N_2) inerting gas)	12.5	12.0	13.8	12.4
Minimum ignition temperature (°C)	420.0	400.0	610.0	450.0

(Source: Mutanen, Nissinen and Linna, 1990)

Spraying trucks or railway cars with latex sealer may help in controlling spontaneous combustion by keeping dust down during lignite transport (Couch, 1990a). In Finland, storage bins for milled peat at one plant are lined with a fireproof veneer, equipped with fire and explosion vents and fire detectors. Equipment for fighting fire with water or foam is also kept nearby (Kalmari, Maskuniitty and Kosunen, 1990). Compaction of the material may also reduce the risk of spontaneous combustion or explosion by decreasing the incidence of air pockets.

Impacts to the physical environment due to the transport of low-grade fuels

The transport of low-grade fuels may have an impact on soil, air and water quality primarily from dust escaping en route or as a result of spills during loading or unloading or along transport routes. Lignite or oil shale dust can settle on agricultural lands or bodies of surface water, potentially disrupting fish or wildlife habitat and contaminating soils. Choosing the optimal transport route (e.g. avoiding densely populated areas or fertile agricultural lands) and preventing the dust from leaving the fuel carrier through the use of closely fitted covers, the application of sealants or heavy oil (e.g. 7–9 l/t) can help to reduce the quantity and magnitude of these potential effects (Couch, 1990a; Dholakia and Bachelor, 1988). Driving at slower speeds will also aid in minimizing the escape of dust. A summary of potential impacts and proposed control measures is presented in Table 5.2.

Disruption to residents along transport routes

People living along routes where low-grade fuels are transported can experience disruption due to increased levels of noise, dust and traffic from both trucks and trains. Residents adjacent to roadways will also be exposed to higher levels of truck traffic and will, therefore, face a greater risk of traffic-related injury or death. A greater number of low grade fuel-carrying freight trains can also increase the chance of accidents at level railway crossings.

Effects such as those outlined above can be at least partially mitigated through driver training for truck drivers carrying low-grade fuels. An assessment of the best transport route based on traffic levels and location of sensitive features such as schools and hospitals along proposed routes, as well as transporting when traffic is lighter, will help to minimize risks.

Table 5.2 *Impacts to the physical environment during low-grade fuel transport*

Potential impact	Environmental protection measure
Decrease in quality of the physical environment	
■loss of soil quality	■plan transportation routes away from existing agricultural land use; cover or contain fuel during transport where possible
■loss of water quality	■monitor water quality along transport routes and re-route if water quality is declining; cover or contain fuel during transport wherever possible
■loss of air quality	■monitor prevailing wind conditions and assess particulate matter in ambient air along potential routes; choose transportation routes away from sensitive features such as schools and hospitals; avoid operating at high speeds, particularly during windy conditions; cover or contain fuel during transport whenever possible
■increased risk of upset or spill	■develop remedial action plan for use in event of spillage; move quickly to retrieve fuel and minimize impact on air, soil and water

Operating during daylight hours will reduce the nuisance effects of noise and vibration. Dust along transport routes may be controlled by not overloading transport containers, driving at lower speeds and by using covers or contained carriers where possible (Table 5.3).

Health risks of low-grade fuel transport

The possibility of a road accident poses the greatest risk of death or injury during the transport of low-grade fuels. Table 5.4 presents information regarding health risks associated with the transport of coal, peat and oil shale. Overall, rail is currently the safest option. Fatalities and injuries resulting from the road transport of peat are higher than for coal, perhaps due to the greater number of trucks required to transport peat (generally a lower calorific value fuel than coal) per unit of energy produced. Information was not available regarding the type or rank of coal to which these figures apply.

Table 5.3 *Potential disruption to residents along low-grade fuel transport routes*

Potential impact	Environmental protection measure
Disruption to residents along transportation routes	
■increased risk of accident or injury	■plan transport route along areas of lowest population density and least sensitive existing land use; transport during least disruptive times (e.g. avoid rush hours), using experienced and well-trained drivers
■increased nuisance effects from noise, dust and vibration	■operate during daytime hours; cover and contain fuel where possible

Table 5.4 *Health risks of fuel transport (data in fatalities/GW(e))*

Fuel	Mode	Nature of risk	Occupational	Public
Coal	Rail	*	0.005–0.022[a], 0.00015[b]	0.37–0.54[a]
	Truck	immediate	0.6–24[a], 0.014–1.23[c]	0.6–24.0[a]
	Barge	*	0.25–1.0[a], 0.062[b]	0.07–0.28[a]
	Pipeline	*	0.16[a]	
Peat	Truck	immediate	0.35[b,d]	0.9–14.0[e]
Oil shale	Rail	immediate	0.02[e,f]	0.05–0.07[c,g]

[a] Hamilton, 1984
[b] Hubert et al., 1981
[c] UNEP, 1985
[d] Approximate figure, estimated from coal and fuelwood figures
[e] Hamilton, 1990
[f] IWG Corporation, 1982
[g] House, 1983

* not evaluated

(Source: Compiled from IAEA, 1991)

Worker health and safety during low-grade fuel transport

Research into occupational health and safety issues relating to the transport of coal has indicated that workers can suffer effects from dust during loading and unloading and accidents with loading and unloading equipment. These impacts would be relevant for individuals working with oil shale, lignite and peat, particularly if large-scale operations such as power production are undertaken. Safety measures would include use of protective filter masks, maintenance of equipment and high levels of worker training in equipment use.

The potential for injury due to the fires or explosions involving peat or lignite means that individuals working with these fuels should be operating under conditions where fire detectors are installed, proper handling and storage facilities are used (e.g. thick-walled containers with explosion vents, cleaning working area to control the spread of any fire which might occur) and fire-fighting equipment is in good working order and can be easily accessed. These recommendations are summarized in Table 5.5.

Table 5.5 *Worker health and safety during low-grade fuel transport*

Potential impact	Environmental protection measure
Worker health and safety	
■risk of injury during loading and unloading	■loading and unloading should be carried out following a strict safety code by well-trained personnel; high-quality filter masks should be worn to reduce exposure to dust; training in safety measures regarding potential ignition or explosion of lignite or peat should be provided

References

CHADWICK, M J, HIGHTON, N H and PALMER, J P (1987) *Mining Projects in Developing Countries: A Manual.* Beijer Institute Centre for Resource Assessment and Management, York and Stockholm.

COUCH, G R (1990a) *Lignite Upgrading.* IEA Coal Research, London.

COUCH, G R (1990b) 'Lignite, low-grade coals and peat: an overview of resources, power generating and upgrading'. In *Low-grade Fuels.* Vol 1. Technical Research Centre of Finland, Espoo.

DHOLAKIA, V and BACHELOR, F W (1988) *Detailed Study of the Reaction Mechanism of Spontaneous Combustion of Coal with Particular Emphasis on Western Canada Low-rank Coals, Phase II*. CANMET, Alberta.

HAMILTON, L D (1992) 'Health and environmental management of low-grade fuels'. In *The Environmentally Sound Management of Low-grade Fuels*. SEI, Stockholm.

HAMILTON, L D (1990) *Personal Communication*. Brookhaven National Laboratory, Upton.

HAMILTON, L D (1984) 'Health and environmental risks of energy systems'. *Risks and Benefits of Energy Systems*. IAEA, Vienna.

HAMILTON, L D (1979) 'Health effects of electricity generation'. *Symposium on Health Effects of Energy Production*. Chalk River Nuclear Laboratories, Ontario.

HOUSE, P W (1983) *Report EP-0093*. US Department of Energy, Washington.

HUBERT, P, MOATTI, J P, MACCIA, C and FAGNANI, F (1981) *Les Impacts Sanitaires et Ecologiques de la Production d'Electricité: le cas français*. Centre d'étude sur l'évaluation de la protection dans le domaine nucléair, Fontenay-aux-Roses.

HUETTENHAIN, H (1992) 'Low-rank coal upgrading technology review'. *Conference on Clean and Efficient Use of Coal: A New Era for Low-rank Coal*. IEA/OECD, Paris.

IAEA (1991) 'Comparative environmental and health effects of different energy systems for electricity generation'. In *Senior Expert Symposium on Electricity and the Environment: Key Issue Papers*. IAEA, Vienna.

IWG Corporation (1982) *Health and Environmental Effects: Document on Oil Shale*. IWG, San Diego.

KALMARI, A, MASKUNIITTY, H and KOSUNEN, P (1990) 'Handling systems for peat, wood and sludge in industrial projects'. In *Low-grade Fuels*. Vol 2. Technical Research Centre of Finland, Espoo.

LAINE, R (1983) 'A study on the storage of stump chips and sod peat in a shed'. *Turve Teollisuus* 4: 175–190 (in Finnish, English summary).

LISHTVAN, I I, GAVRILCHIK, A P and FALYUSHIN, P L (1985) 'Self-heating and self-ignition of peat during storage'. In *Proceedings of International Peat Society Symposium 1985*. Joenkoeping, Sweden.

MORRIS, S C, NOVAK, K M and HAMILTON, L D (1979) *Databook for the Quantification of Health Effects from Coal Energy Systems*. Biomedical and Environmental Assessment Division, Upton.

MUTANEN, K, NISSINEN, K and LINNA, V (1990) 'Improvement of safety in peat handling'. In *Low-grade Fuels*. Vol 2. Technical Research Centre of Finland, Espoo.

SZABO, M F (1978) *Environmental Assessment of Coal Transportation*. PEDCO Environmental, Cincinnati.

UNEP (1985) *The Environmental Impacts of the Production and Use of Energy*. Part 4. United Nations Environment Programme, Nairobi.

6

Environmental management during processing and upgrading of low-grade fuels

Low-grade fuels are upgraded and processed in many ways. Procedures are carried out to achieve a cleaner burning fuel, to increase the calorific value of the fuel by weight, as an interim step in the production of chemical feedstocks or as preparation for use in industry or power production. Fuels may be upgraded to liquid or gaseous form, or they may be compressed to produce a fuel with a greater bulk density.

During low-grade fuel processing or upgrading the characteristics of the fuel are altered, and in most cases, this results in an increased calorific value due to the removal of significant amounts of water and inorganic matter. This can have potential impacts on the environment through the production of large quantities of effluent, steam condensate and solid waste which requires disposal and through the application of energy to carry out the process. The major environmental impacts of processing and upgrading low-grade fuels are discussed below.

Identifying and implementing the best environmental management strategies require a systematic assessment of the potential impacts resulting from an undertaking such as fuel processing and upgrading. Chapter 3 provides a summary of the level of detail required to make an informed decision regarding the best practice and technology for environmental protection during oil shale surface retorting. The approach discussed, however, is exemplary and should form part of pre-operational environmental planning for any fuel technology.

Peat processing

The majority of peat processing occurs using milled peat, both since peat is more frequently harvested in milled form and because wet mining produces milled peat which is too waterlogged to allow combustion without some form of treatment. Sod peat, although sometimes used for coking, generally undergoes little or no processing prior to conversion, since during harvest the peatland is drained and the cut sods are dried by the sun, resulting in a fuel with a moisture content suitable for burning. The main effects from peat processing and upgrading are presented in Table 6.1.

Table 6.1 *Environmental impacts of peat processing*

Potential impact	Environmental protection measure
Disruption of the physical environment	
■escape of solvents to groundwater and nearby waters	■contain and recycle solvents used for peat dewatering; use settling ponds and/or surface addition of chemicals to prevent solvent escape
■combustion of energy to dry peat	■use waste from peat production or other low-sulphur fuels to prevent a decrease in air quality and a build-up of solid waste

Peat drying and dewatering

Solar and air drying Depending upon its moisture content and the prevailing weather conditions such as temperature and precipitation, peat is solar-dried in situ. Very wet peat will be spread over a nearby field and left to dry. If it is not possible to obtain the desired moisture level for burning through air drying to at least 45–55 per cent (Boron, Evans and Peterson, 1987; Bord na Móna, 1985; Robertson and Godsman, 1975), mechanical methods for pressing and squeezing the peat, sometimes in conjunction with the use of solvents, will be employed. Thermal heat may also be applied in the manufacture of peat pellets for power production. Steps undertaken to achieve a dry peat with a high calorific value (20–30 MJ/kg) are presented in Figure 6.1.

Mechanical pressing and solvents Mechanical pressing involves the use of rollers, filters and screens to express excess moisture. Pressing alone rarely reduces the moisture content to much less than 65–80 per cent. To release higher quantities of moisture, solvents such as diethylketone or benzene must be added prior to processing to break down the material structure of the peat.

The use of solvents can have a serious impact on water quality if they are carried to adjacent surface waters or allowed to escape into the groundwater under the processing site. This potential problem can be avoided by containing all solvent-laden waters or recovering and treating all solvents prior to disposal. The use of solvents is expensive, and containment and recycling frequently occur in order to reduce the costs of their use.

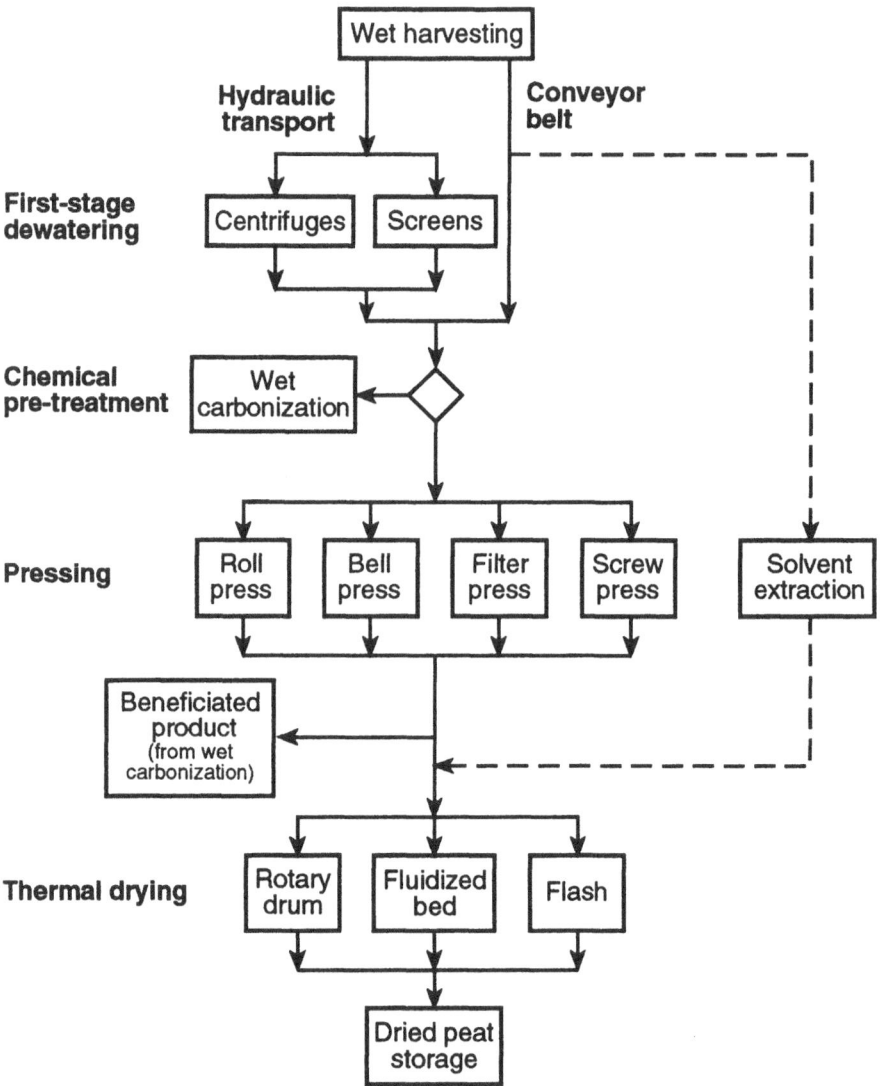

Figure 6.1 *Alternative technical methods of peat dewatering*

(Source: Tsaros, 1981)

Thermal heat Thermal heat can be used along with chemicals to reduce the moisture content of peat. An external source of heat applied to wet peat is known as wet carbonization. Wet oxidation occurs through partial combustion of the peat. Both of these processes are followed by mechanical pressing. Wet carbonization and wet oxidation are high-cost procedures still primarily in the experimental stage.

Carbonization

Sod peat is the usual raw material for the carbonization or coking of peat. Peat used for carbonization requires a moisture content of 35–45 per cent, and when coking is carried out, the peat is screened and fed into a carbonizing kiln. Ideally, the heat applied to the kiln will come from peat fines or peat dust that might otherwise be lost. Carbonization of peat produces coke which is lower in impurities such as sulphur and phosphorus than most lignite or higher-grade coals.

Compaction

When peat is harvested by mechanical methods, much of the dried material is in the form of small detached particles known as fines. Although this physical form may be suitable and, in many cases, preferable for use in large boilers, it is not readily usable in small boilers or domestic heating appliances. To be used in this manner, peat fines must be compacted and formed into larger pieces which may be handled more easily. In Australia, compaction technology for brown coal has been used to simplify the transport of peat fines to power stations and to maintain combustion stability in the boiler.

Compaction does not alter either the density or the inherent physical or combustion properties of the peat, though its moisture content may be reduced if intermediate heating is used (Eriksson and Prior, 1990). Other forms of solid fuel fines, including lignite and bituminous coals as well as charcoal fines, may be compacted in a similar way. The resultant product is usually called a briquette, though if produced by a roller press similar to that used for making cattle feed, the small cylindrical product is often called a pellet.

There are a number of mechanical machines which have been developed to achieve this compaction; essentially all rely upon application of moderate pressure (say up to 50 million Pascal (MPa)) and, usually, upon the use of a binding agent to hold the fine particles together. A number of such binding agents have been used, including lime, clay, molasses and various types of organic gum. The most common binder in modern high-speed machines producing lignite or anthracite briquettes (the most common type of fossil fuel briquette) is ligno-sulphate produced as a by-product of paper manufacture.

Binderless briquettes can be made from certain types of initial feed. In these processes the material is usually heated to approximately 65°C, which promotes a softening of the particles, and subjected to pressures up to 150 MPa.

The most common machine used to make briquettes in Europe is a high-speed roller or multi-piston press. In China, where briquettes made from bituminous coals or anthracite are very common as a household fuel, they are made in a much less mechanized fashion in a batch piston press. In both cases, however, the product is essentially the same, varying only in the properties of the original fuel.

The compacting process neither produces any effluent stream nor appreciably alters the environmental impacts of burning the original fuel. Since compaction enables the use of fines, which would otherwise be left as wasted energy, it may have a slightly beneficial impact on environmental quality.

Lignite upgrading

Most lignite upgrading is carried out to improve the operation of a power station, although processed lignite has other possible uses as outlined in Figure 6.2. Procedures may be aimed at improving the quality of the lignite so that it may be substituted for higher-ranked coals, evening out the variations in fuel quality and reducing fouling problems and high maintenance costs of boilers.

The choice of upgrading technique is site specific and depends on a number of factors including proposed end use (e.g. power generation in a conventional pulverized fuel system, fluidized bed combustor or integrated gasification combined cycle (IGCC)), size of site available for various procedures and distance between mine and conversion site. The variability in the composition of lignite means that, in most cases, it is not possible to use conventional coal-processing methods without major modifications (Sondreal, 1992; Couch, 1990; BRCA, 1987).

Approaches to lignite upgrading can include homogenization (mixing), cleaning (removal of mineral matter and pyritic sulphur) and thermal drying (moisture removal), the latter often taking place under heat and pressure. Huettenhain (1992) has identified six levels of upgrading for low-rank coals which are presented in Table 6.2 along with an assessment of the effect of upgrading on the fuel and boiler operating conditions.

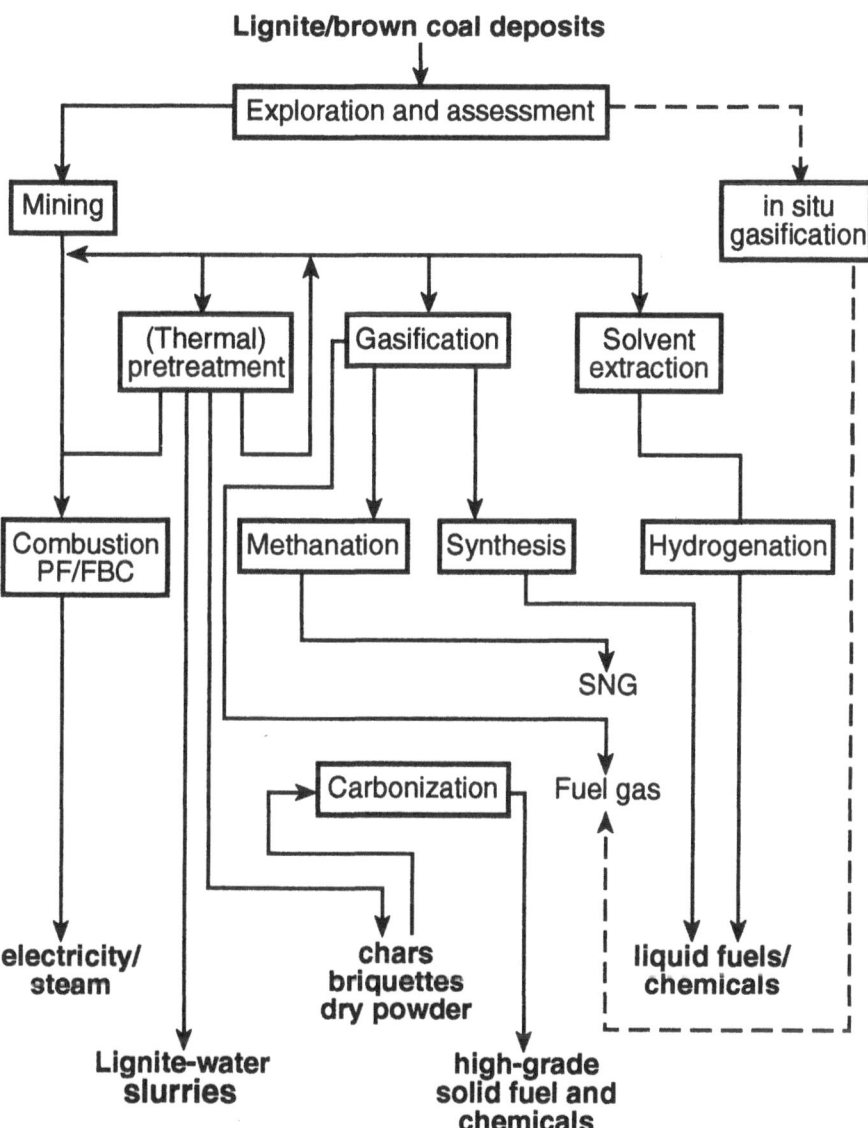

Figure 6.2 *Options for lignite utilization*

(Source: Couch, 1990)

Table 6.2 *Low-rank coal upgrading technology comparison*

Technology	Fuel impact	Boiler benefit	
		Existing boiler	New boiler
No upgrading	Base	Base	Base
Level 1 Mixing	Less quality variation	Less derating	Tighter design
Level 2 Rock removal and/or coarse coal cleaning	Lower ash content	Lower maintenance	Lower maintenance
Level 3 Fuel cleaning	Less ash and sulphur	Higher efficiency, lower emissions	Higher efficiency, lower emissions
Level 4 Thermal drying	Higher heating value	Less derating, higher efficiency	Higher efficiency, lower emissions
Level 5 Coarse coal steam drying	Stable transportable and higher heating value fuel	Replace bituminous coal	Smaller boiler
Level 6 Hot water-drying new fuel forms	Lower sodium, sulphur, ash	Replace bituminous coal	Much smaller boiler

(Source: Huettenhain, 1992)

Homogenization

The variability within a deposit may be reduced by mixing the mined lignite prior to combustion. This requires large areas for storage and manipulation by heavy equipment. Homogenization will decrease the variability in the fuel, placing less demand on the boiler with respect to the range in fuel components such as moisture and mineral matter content and, therefore, potentially increasing the efficiency and life span of the power generating system.

Cleaning

Cleaning lignite involves physically crushing and screening the fuel to remove quantities of inorganics. Chemical cleaning (e.g. leaching with an aqueous nitric acid) may be used in conjunction with physical cleaning. Advanced lignite washing techniques capable of removing up to 60–82 per cent of ash prior to combustion are also being developed. An example of this originates from the Energy and Environment Research Centre (EERC) at the University of North Dakota, where the lignite is 'sized to minus 30 mesh, acid leached, treated with a surfactant and then agglomerated with a phenolic oil (Sondreal, 1992)'. All cleaning techniques for removing mineral matter produce solid waste which requires careful disposal.

Moisture removal and thermal upgrading

Lignite is 35–70 per cent moisture as mined. Moisture is removed from lignite by mechanical pressing (for small-scale and non-commercial use), evaporation and by a combination of heat and pressure. Three of the four best understood methods for dewatering lignite make use of conventional evaporative drying.

Water is held in lignite in four ways, including unbonded, in capillary form, in a physically absorbed multimolecular layer and in a monolayer which is at least partially chemisorbed (BCRA, 1987). The application of heat in various forms and at a variety of temperatures and pressures will break down the water-holding mechanisms of lignite and allow the moisture to escape in steam condensate or liquid form.

Lignite Upgrading (Couch, 1990) provides the most up-to-date and comprehensive review of methods for processing lignite and is the source of much of the technical information provided in this section. Table 6.3 lists all the processes described in detail by Couch (1990). The vast majority of these procedures are in the development or experimental stage principally due to costs and engineering problems associated with the wide range of fuel characteristics inherent within and among lignite deposits. Methods that are considered to be 'well established' are outlined below as well as a discussion of the environmental effects that should be anticipated during lignite upgrading. Some techniques which are not yet at the commercial stage (e.g. lignite drying using a fluidized bed) show greatest promise at the low end of the moisture range (Couch, 1990).

Table 6.3 *Summary of processes for lignite drying and upgrading*

Process	Nature	Status	Product	Comment
Flash mill drying	evaporative	well established	dried and finely divided	■ an integral part of utility boiler operation; could operate on a stand-alone basis
Rotary drum drying	evaporative	well established	<12 mm	■ not currently used because of explosion hazard
Steam tube drying	evaporative	well established	<6 mm	■ commonly used process, especially in conjunction with briquetting process
Fluidized bed drying	evaporative	pilot/ demonstration	dried and fine	■ single-stage demonstration plant in GDR; bench-scale testing of multiple-stage process
Swirl tube drying	evaporative	demonstration	fine powder	■ used with indirect heating to obtain low moisture content down to 2 per cent; mechanical design is difficult
Drying and oil treatment	evaporative	pilot	depends on feed	■ the oil treatment is to stabilize the dried material for transport; production unit on sub-bituminous coal being commissioned in the USA
Solar drying	fine milling and evaporative	pilot	lumpy and crushed to <150 mm	■ dependent on climate and land available; work done in Australia

Table 6.3 (continued)

Process	Nature	Status	Product	Comment
Densification	fine milling and evaporative use of physical force	lab scale	extruded pellets a mixture	■ interesting new development from Australia
Press dewatering		small scale		■ complex mechanical process requiring 2 to 3 stages to achieve 15 per cent moisture
Batch Fleissner	thermal upgrading	well established	lump	■ long established commercial process, relatively complex batch cycle and high cost
Continuous Fleissner	thermal upgrading	pilot	small lumps or milled	■ relatively complex mechanical design powder needed with possible wear problems
Evans-Siemon (water)	thermal upgrading	lab scale	lumps	■ a semi-continuous version of the Fleissner process
UNDMRC (hot water)	thermal upgrading	pilot	lignite-water slurry	■ thoroughly tested process produces a stable slurry; the lignite can be slurried in its own moisture
UNDMRC (steam)	thermal upgrading	bench scale	lumps	■ only of interest if a lump is needed
Bechtel (hot water)	thermal upgrading	pilot	slurry	■ pilot work carried out on test sub-bituminous

Table 6.3 (continued)

Process	Nature	Status	Product	Comment
SPC (steam)	thermal upgrading	pilot	fine powder	
IGT (hot water)	thermal upgrading	lab scale	lumps	
Koppelman	thermal	pilot	pellets	■a demonstration unit is being built for Wyoming sub-bituminous coal; a great deal of pilot plant development work has been undertaken; mechanically complex but produces a high-grade product ■little published data
WEECO	drying and partial pyrolysis	pilot	powder	
Hitachi	drying and partial pyrolysis	pilot	no data	
Mitsubishi	drying and partial pyrolysis	pilot	powder	■fluidized bed based
DK	thermal	lab scale	lumps	■a version of the fluidized process
Thermocoal	thermal upgrading	pilot	powder or briquettes	■developed for possible use with the USSR

Flash drying The most widely used upgrading technique is flash drying, which often forms part of the power production process where recycled hot flue gases can be used to dry the lignite (Figure 6.3). In this case, the recovered moisture is in the form of steam condensate which will be passed through an electrostatic precipitator to remove particulates (Couch, 1990). Flash drying may also take place independently from power production, but this requires the combustion of fuel specifically for drying lignite rather than using waste heat from a generator. Flash drying can accommodate relatively large pieces of lignite.

Steam tube drying In this process, crushed lignite is rotated in a large shell-like structure through which hot steam is passed. This technique is most often used to prepare lignite for briquette production. Gases leaving the shell are passed through an electrostatic precipitator to decrease the potential for a negative impact on air quality.

Rotary drum drying Rotary drum drying involves placing relatively large fragments of lignite into a rotating tube. The lignite is then dried using flue gas from power generation. The presence of highly combustible lignite dust in the drum can lead to explosion, a major factor in explaining why this technique is seldom used today.

Drying and thermal upgrading

Fleissner process The Fleissner process uses lignite pieces at least 10 mm square in size which are then heated under pressure in the presence of steam (Figure 6.4). This procedure produces a higher calorific value lignite and a liquid containing oils, tars, ammonia, sulphur compounds, alkali metals and organics.

The process has been traditionally carried out in batches where energy efficiency is improved by releasing the hot steam from one drying chamber to the next. Experiments are ongoing in an effort to achieve continuous drying.

Briquetting

For details of lignite briquetting see above.

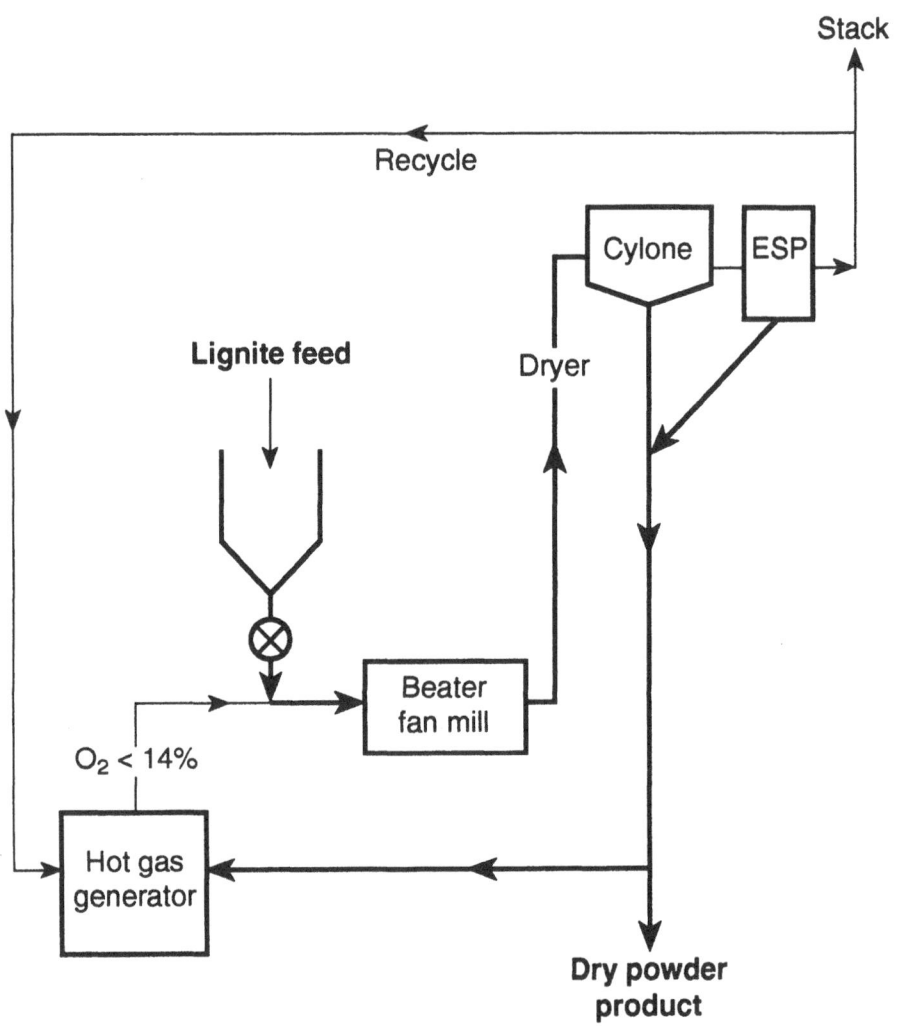

Figure 6.3 *Flash-drying circuit*

(Source: Couch, 1990)

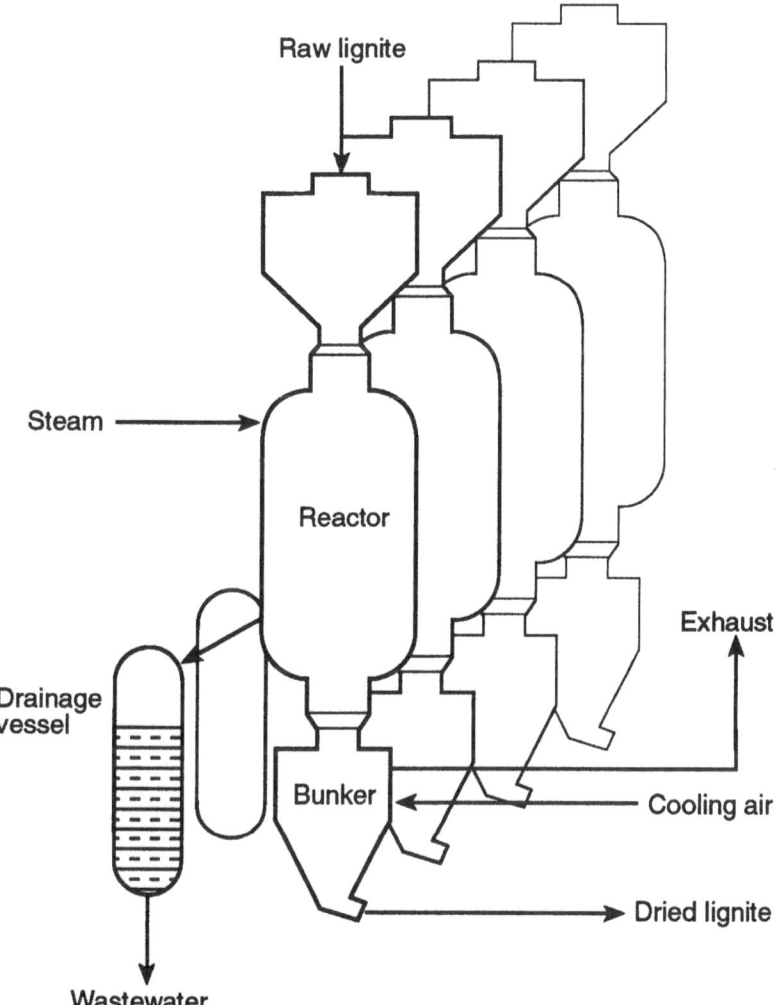

Figure 6.4 *The Fleissner batch process*

(Source: Fohl and Tessmer, 1985)

Environmental implications of lignite processing and upgrading

The processing and upgrading of lignite pose the greatest threat to environmental quality through the production of effluent and/or solid waste. The discussion presented below is summarized in Table 6.4.

Table 6.4 *Potential environmental impacts resulting from lignite processing and upgrading*

Potential impact	Environmental protection measure
Impact on the physical environment	
■production of wastewater containing impurities such as phenols, tars, oils, sodium compounds and sulphur compounds	■closed system for containment and treatment of contaminated waters; collection and treatment of wastewater using settling ponds and/or addition of appropriate chemicals (e.g. limestone to hold sulphur) is second best option and requires stringent monitoring to prevent contamination of nearby water bodies and soils
■production of steam condensate containing sulphur and particulate matter	■hot water vapour must be passed through an environmental protection system such as electrostatic precipitators to reduce atmospheric and soil contamination
■build-up of solid waste	■return unwanted mineral matter to mine where possible; if waste contains harmful substances which may be leached to ground and surface waters, these must be removed and carefully stored or treated; it may be possible to revegetate piles of waste mineral matter; cover piles with soil to deter spontaneous combustion; keep mounds stable and low in height to prevent landslides
■combustion of energy	■use waste heat from power production where possible for drying and upgrading purposes (e.g. hot flue gases from power station)

Table 6.4 (continued)

Potential impact	Environmental protection measure
Disruption to residents in the vicinity of the processing site	
■visual impact due to visible plume	■use of electrostatic precipitators will result in cleaner flue gases; inform residents about contents of plume to decrease concerns regarding impact of materials released into the atmosphere
■build-up of solid waste	■avoid build-up of waste where possible (e.g. return immediately to mine); ensure harmful constituents of waste are removed or contained to prevent contamination of groundwater which may serve as drinking water for area residents; revegetation of disposal areas where possible

Steam condensate The production of steam condensate is a primary environmental consideration during lignite upgrading. The hot steam will contain sulphur and particulates which must be removed prior to release into the atmosphere. This is especially important during the use of high-sulphur lignites. Electrostatic precipitators are the most common mitigating measure employed to protect air quality during flash drying. Hot water vapour may form a heavy white plume which will be visible across the landscape.

Liquid effluent Liquid effluent from the Fleissner process will require containment and/or treatment to prevent polluting substances such as tars, phenols and sulphur compounds escaping into the environment. Depending on the disposal facilities available, the production of these materials can contribute to the build-up of toxic waste. If the waters are retained, it may be possible to separate some of the substances such as humic acids and phenols for sale as chemical feedstocks. Simply disposing of effluent from lignite processing in nearby water bodies or lands can have a severe negative impact on surface and groundwater quality, soil fertility, plant, fish and wildlife habitat and generally poison and disrupt any associated ecosystem.

Solid waste Beneficiation will lead to the production of large quantities of the mineral matter removed from the lignite. This can have a negative impact on soil and water quality, particularly during periods of rainfall when leaching may occur. The aesthetic nature of the landscape may be affected through the presence of heaps adjacent to beneficiation activities. In many cases it may be possible to return the extracted materials to the mine or revegetate the areas where the ash is stored. The waste should not be piled too high or steeply to guard against the possibility of a landslide.

Lignite is a highly combustible fuel, and problems of spontaneous combustion can arise in waste heaps from lignite processing. A burning pile will release sulphur and particulate-laden gas into the atmosphere and can also prevent successful revegetation by burning small seedlings or creating soil temperatures that are too high for germination to occur. Waste of this type should be well covered with soil to decrease the availability of oxygen to the lignite fines. An area which is smoking or glowing should also be immediately smothered.

Oil shale processing and upgrading

Oil shale is a solid fuel which can be burned 'as mined', or it may undergo treatment to yield a liquid fuel known as shale oil. There are two principal methods for extracting the carbon-rich kerogen from oil shale – surface retorting (sometimes known as ex situ retorting) and in situ retorting. The fuel that is produced through retorting frequently requires refining prior to combustion. The costs of refining are generally high and have been estimated to be approximately 20–40 per cent of the total cost of producing oil from oil shale (Ekinci, 1995).

The complex chemical character of oil shale leads to the formation or release of a number of potentially toxic and polluting compounds during processing and upgrading, including phenols, polycyclic aromatic hydrocarbons, recorsin derivatives, quinones, carbonic acids and heavy metals (Lee, 1991; Lurgi, 1988; Aarna, 1978; Yen, 1976; Yen and Chilingarian, 1976). It is important to emphasize the potentially serious damage which these substances can have on existing ecosystems, either independently or in a cumulative fashion, if they are not carefully controlled or treated.

There is also a risk to the health and safety of oil shale workers during crushing, retorting, upgrading and refining. The principal risk is from exposure to air pollutants such as carcinogenic heavy metals (e.g. arsenic, cadmium, chromium and nickel), polycyclic aromatic hydrocarbons and radionuclides which may be released during retorting (Hamilton, 1990; House, 1983, 1981; IWG Corporation, 1982).

The impact of processing can be particularly harsh for nearby residents. Oil shale development will almost certainly take place in rural areas where blasting and crushing operations will be heard against a background of low ambient noise levels.

Surface retorting

As Figure 6.5 illustrates, surface retorting of oil shale comprises a series of steps, each of which provides an opportunity for the production of a variety of contaminants. An environmental characterization information report for surface retorting of oil shale is presented below.

Dust The first stage is the crushing of mined oil shale fragments into pieces 5–75 mm square in size. Pulverizing of the raw oil shale can produce large quantities of dust, which may be partially controlled through the use of filters in the crusher units (Hamarneh, 1985). Some of the fines produced during crushing may be burned to provide partial heat for the retort or they may be mixed with spent shale for disposal (Attassi, 1992). Wetting agents and water may be used to control dust. Water from retorting, however, is unsuitable for this purpose, since it is full of contaminants.

Gases The breakdown of oil shale into crude shale oil, retort gas and spent shale takes place in a retort. The heat for retorting may be provided directly (e.g. hot retort gas to the solid oil shale) or indirectly (transferred from a separate furnace outside the retort (UNEP, 1985)). Atmospheric pressure fluidized bed retorting of oil shale has been found to increase the yield of oil from oil shale over other technologies for the Devonian shales of the eastern United States (Carter et al., 1995). During retorting, semicoking of some of the oil shale may occur (Purre and Vasilyev, 1992). The semicoke will often be burned along with the retort gases to provide a source of heat for the retort.

A quantity of low-calorific gas may also be created during high-temperature processing. Following the removal of hydrogen sulphide and ammonia (EPA, 1981), this gas can be burned in specially designed boilers fuelled by waste heat (Zhiryakov, Ryatsep and Aunap, 1992).

Surface and in situ retorting can result in emissions of sulphur and nitrogen oxides, CO, CO_2, light hydrocarbon vapours (lower than from petrochemical plants) (Zhiryakov, Ryatsep and Aunap, 1992) and other organic compounds. Sulphur and larger particulate matter can be controlled through the use of flue gas cleaning (e.g. electrostatic precipitators, scrubbing, baghouses), although these methods will not significantly reduce the amount of CO_2 produced.

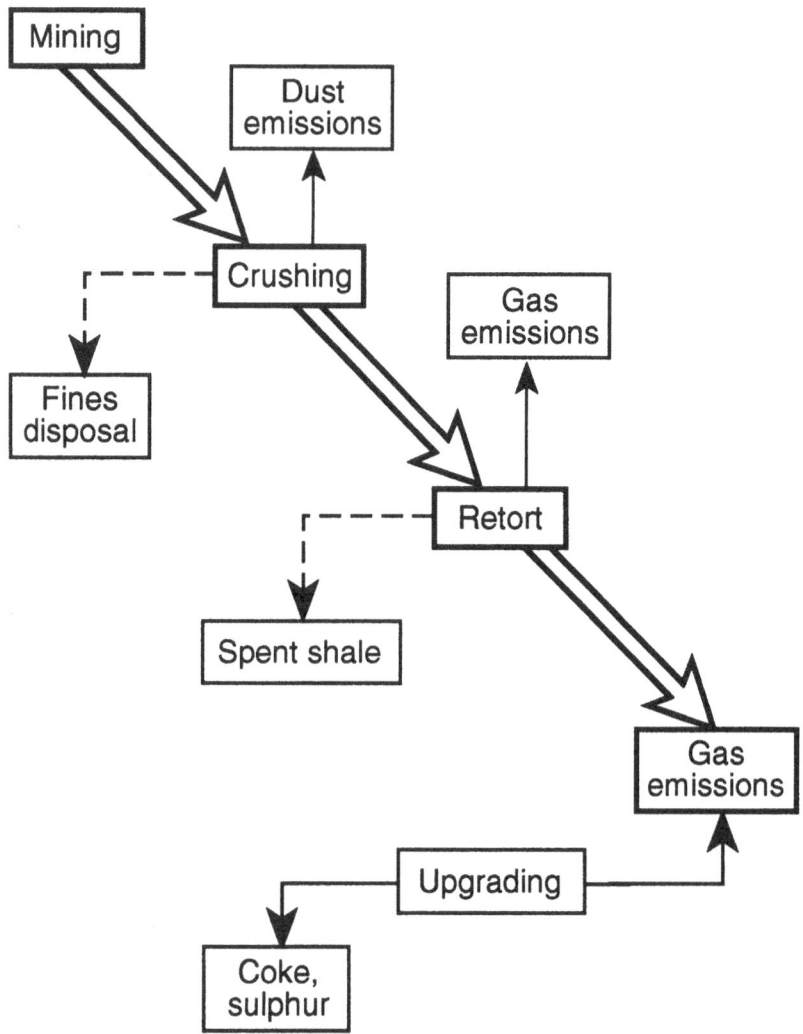

Figure 6.5 *Flow diagram for surface retorting of oil shales*

(Source: UNEP, 1985)

Solid waste Surface retorting will produce great quantities of waste material in the form of spent shale (Groppo, 1995) and fly ash caught in the electrostatic precipitators. The wide variation in the composition of ash from oil shale makes it particularly difficult to use, so it is often stockpiled (in mounds up to 100 m high) or returned to the mine.

Oil shale waste is combustible, and waste heaps will burn for long periods of time if a fire begins. In Kumla, Sweden, a large waste heap left over from oil shale processing has been burning since the Second World War. Steps to reduce the likelihood of burning include enforcement of a strict policy of no smoking or lighting of fires in the vicinity of the mounds and immediate smothering of any areas which are showing signs of burning. If a fire does start it is more easily controlled if the waste is stored in several smaller piles as opposed to one large one.

Stockpiling or returning waste to the mine leaves the spent shale susceptible to wind and water erosion. Since oil shale dust may contain traces of heavy metals such as vanadium or molybdenum and toxic compounds (e.g. phenols) which can contaminate land and water bodies, its disposal must be carefully controlled. Estonia's major waste problem concerns the indiscriminate disposal of its hazardous waste, which comes almost entirely from oil shale (Pierce, 1993).

The mining site should be prepared prior to the return of any spent shale. In particular, spent shale should not be returned to a wet mine, as this encourages leaching (US Departmernt of Energy, 1988). Environmental protection measures would include lining the site with clay, laying down leachate monitors, designing leachate collection sites at the bottom of the area to be filled and installing pumps. These measures are particularly important where the disposal areas interact with flows of groundwater or are located near a major aquifer. Combustion ash must also be layered with the overburden removed during mining to stabilize the fill and help prevent the escape of pollutants.

Areas where spent shale is to be stored should also receive treatment similar to that outlined above.

In situ retorting

The in situ retorting of oil shale takes place in two basic ways. True in situ (TIS) retorting, where hot steam and air are forced into an oil shale deposit still in the ground, involves no mining. With modified in situ (MIS) retorting, the overburden is removed and charges are placed, then detonated, throughout a block of oil shale. The top layer of fragmented oil shale is ignited, and steam and air are forced into the retorting area. This sets up conditions of pyrolysis which cause the release of shale oil. The shale oil along with retorting gases and water vapour will either rise due to pressure from the overlying mass or be pumped to the surface. Modified in situ retorting is illustrated in Figure 6.6.

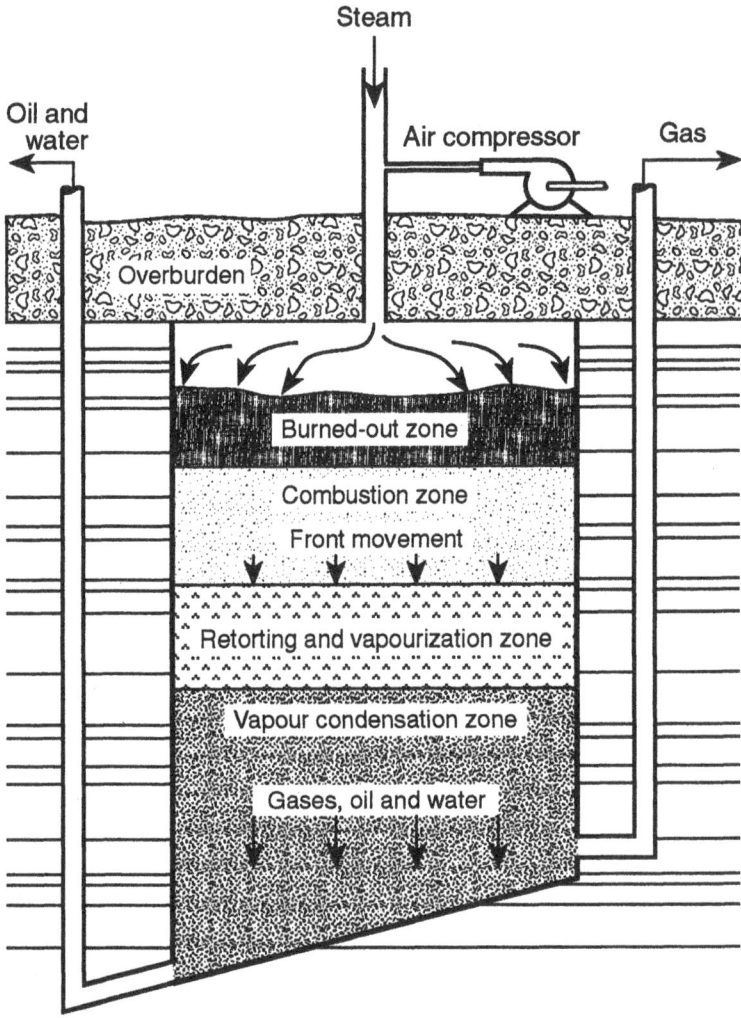

Figure 6.6 *Schematic diagram of MIS retorting (Oxy retort)*

(Source: Probestein and Hicks, 1982)

In situ retorting reduces problems of solid waste disposal associated with surface retorting, although the potential for interaction with and therefore contamination of groundwater is greater (even though in situ retorting uses less than half the water of surface retorting (Penner, 1982)). The nature and severity of other environmental impacts associated with in situ retorting are of the same order as those discussed for surface retorting. More blasting, however, may be required in working the site block by block.

Shale oil upgrading

The shale oil which is retrieved following retorting is extremely dense and viscose at ambient air temperature and requires refining if it is to be shipped by pipeline or used to manufacture liquid fuels. The approach taken to shale oil upgrading is primarily determined by the desired end use (Probstein and Hicks, 1982). During refining there is the potential for pollutant emissions of sulphur, nitrogen and CO_2 as well as the formation of sulphur compounds and heavy tar-containing phenols. Processing to achieve a more refined shale oil will require the removal of arsenic, which must be carefully collected and stored to avoid escape into the surrounding environment.

Water quality

There is great potential for a decline in water quality during in situ and surface oil shale retorting due to large water requirements, removal of groundwater and the potential interaction between the remaining groundwater and spent shale. Treating the wastewater (e.g. removing phenols) may improve water quality but lead to the build-up of toxic or polluting substances, which must be carefully controlled or destroyed.

Evidence from northeast Estonia indicates that the water quality has fallen in two major aquifers due to the production of oil shale. The concentrations of Ca^{++}, $Mg^{''}$, SO_4^{2-} and Cl^- are noticeably higher than in water a greater distance from deposits being mined. It was found that the natural water chemical regime is rapidly restored at the cessation of mining (Erg, 1994).

Figure 6.7 illustrates schematically the circulation of water during oil shale processing at a plant in Estonia, whereas Table 6.5 outlines the contents of the water at each stage. Methods for controlling and eliminating the contamination of water resulting from oil shale processing include:

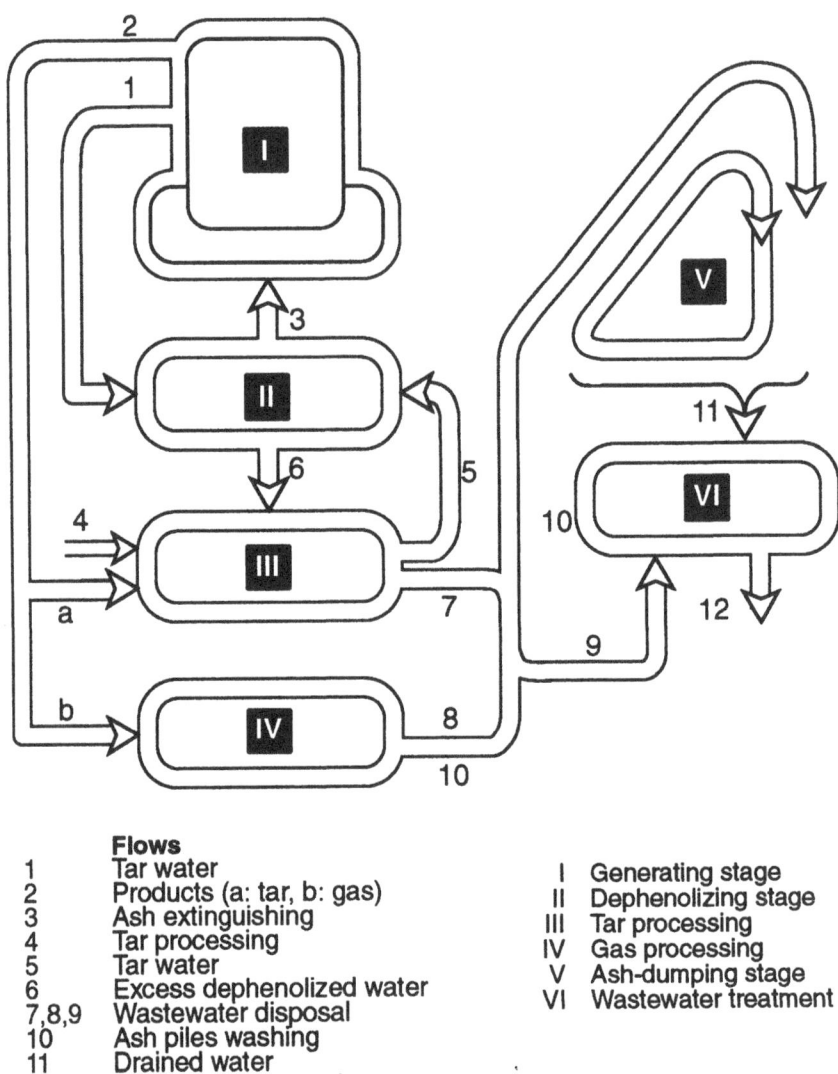

Flows

1	Tar water
2	Products (a: tar, b: gas)
3	Ash extinguishing
4	Tar processing
5	Tar water
6	Excess dephenolized water
7,8,9	Wastewater disposal
10	Ash piles washing
11	Drained water
12	Into natural reservoir

I	Generating stage
II	Dephenolizing stage
III	Tar processing
IV	Gas processing
V	Ash-dumping stage
VI	Wastewater treatment

Figure 6.7 *Water circulation during oil shale processing*

(Source: Zhiryakov, Ryatsep and Aunap, 1992)

Table 6.5 *Composition of wastewater flows (mg·l⁻¹) at different stages of oil shale processing (flow numbers refer to Figure 6.7)*

Parameter	7 and 10	8	9	11	12
			Flows		
pH	6.5	7.1	6.9	12.2	7.6
COD	980.0	1075.0	997.0	792.0	110.0
BOD	580.0	–	602.0	391.0	15.0
Tar	40.0	244.0	46.0	15.5	0.1
Phenols total	233.0	no	200.0	48.5	0.6
volatile phenols	24.0	no	21.0	5.0	traces
aromatic hydro-carbons	209.0	180.0	201.0	–	no
Suspended matter	23.0	39.0	28.0	90.4	20.0
Salts	1200.0	280.0	1005.0	2566.0	635.0
chlorides, Cl^-	236.0	88.0	177.0	381.0	254.0
sulphates, SO_4^{2-}	340.0	84.0	203.0	132.0	–
ammonium, NH_4^+	12.8	–	10.2	3.6	2.0
nitrites, NO_2^-	0.006	–	0.05	–	0.3
nitrates, NO_3^-	0.5	–	0.4	–	1.2

(Source: Zhiryakov, Ryatsep and Aunap, 1992)

- reducing the quantity of water required at each stage of treatment and upgrading;
- using air coolers where cooling is required rather than water cooling;
- improving water-recycling systems (Zhiryakov, Ryatsep and Aunap, 1992); and
- treating phenolic solutions with hydrogen peroxide, in the presence of ferrous sulphate, to aid in the decomposition of phenolic compounds (Preis, 1994).

Environmental management during oil shale processing and upgrading

Oil shale requires considerable treatment to extract and produce a high-quality refined liquid fuel. This processing creates dust, gases, wastewater and solid waste, all of which require treatment and/or containment to prevent contamination of the environment. The discussions presented in the above sections are summarized in Table 6.6.

Table 6.6 *Potential environmental impacts from oil shale processing and upgrading*

Potential impact	Environmental protection measure
Impact on the physical environment	
■dust contamination of nearby surface waters	■spray wetting agents and water during drilling, blasting and crushing; cover conveyors during and land on-site transport of oil shale; use filters in crusher units; pass retort gases through baghouses or electrostatic precipitators to remove particulates
■decline in air quality	■control dust using measures outlined above; clean retort gases by scrubbing, baghouses, electrostatic precipitators; treat gases to remove ammonia
■decline in water quality	■on-site biochemical treatment of wastewater, retrieval of heavy metals and organic compounds; concentration of organic salts into a solid for disposal (Attassi, 1992); prepare mine using clay liner; collect leachate; install monitors and pumps prior to back-filling spent shale and combustion ash; mix ash with overburden before burial
Disruption to residents in the vicinity of the processing plant	
■build-up of solid waste	■return waste to mine following mine preparation; determine quality of ash and recover heavy metals such as nickel and cobalt prior to disposal; increase research into low-waste technologies (e.g. grinding, dehydration and high-temperature oxidation to increase ash quality); a major problem not easily controlled

Table 6.6 (continued)

Potential impact	Environmental protection measure
■noise and vibration	■operate during regular daylight hours; use low-impact explosives where possible; conduct structural survey of buildings adjacent to site and determine appropriate levels of underground vibration; monitor blasting at all times to ensure vibration no greater than 12–15 mm/sec at closest structure; set up community liaison committee to provide point of communication between community and processing plant operators
■risk to worker health and safety	■contain and treat effluents and gases containing heavy metals and organic compounds; ensure proper training of workers, and issue and enforce use of protective wear such as heavy gloves and respirators

Environmental characterization information report for oil shale surface retorting The US Department of Energy (1988) has prepared environmental characterization information reports (ECIR) for several 'hypothetical commercial-sized plants'. The reports present data on the resources required to construct and operate a plant and also provide detail on the types of pollutants and potential environmental disruptions which can occur due to plant operation. This level of detail is essential for choosing the best control technology and preparing an optimum environmental management plan for a given facility.

The ECIR for oil shale surface retorting (Paraho process: direct mode) provides an overview of the process flow chart with environmental points of interest noted (Figure 6.8) as well as an assessment of the resources used through the process, the environmental residuals and by-products produced and gives values for risk of death and injury for workers involved in mining and retorting. These points are summarized and presented in Table 6.7. Mining is included in a consideration of oil shale retorting, since mining must take place so that oil shale is available for processing.

Facility operating parameters assumed for Table 6.7 and Figure 6.8 include:

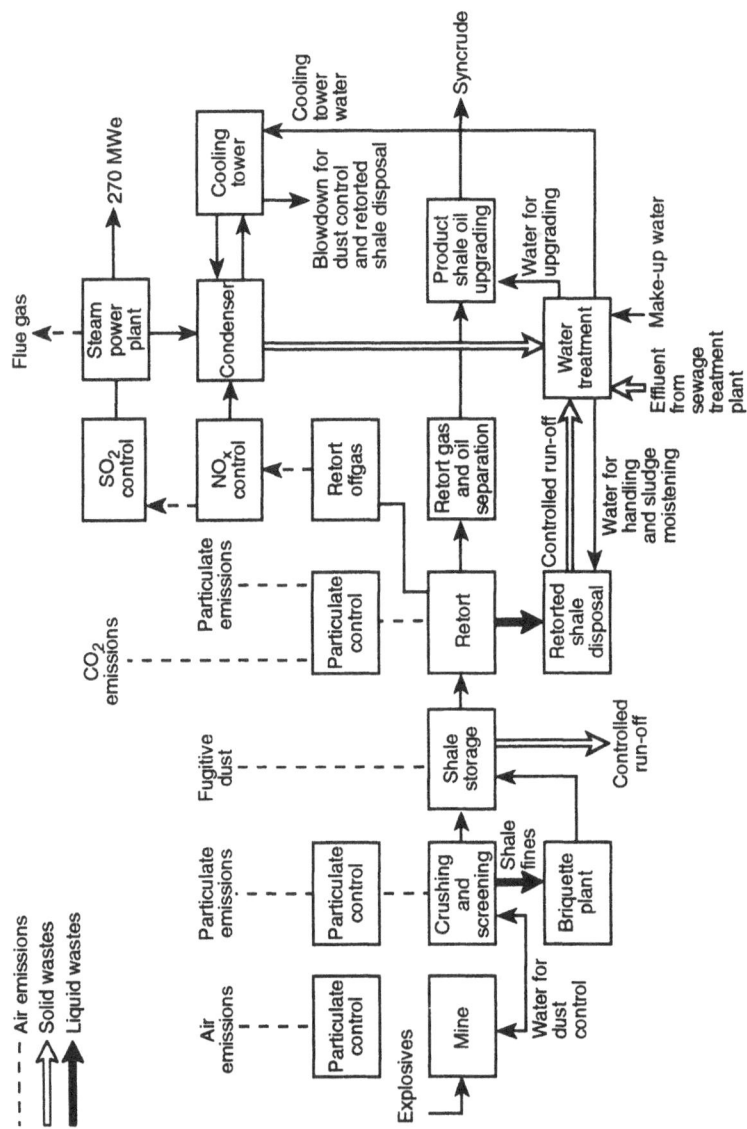

Figure 6.8 *Process flow chart with environmental points of interest – oil shale surface retorting: Paraho process (direct mode) 90 per cent capacity factor*

(Source: US Department of Energy, 1988)

Table 6.7 *Oil shale surface-retorting Paraho process (direct mode)*

Reference energy system

Mine-mouth plant located in the rich oil shale Green River formation; wetting agents, baffled settling chambers, Venturi scrubbers and baghouse filters for particulate control; Phosam-W process for ammonia removal (nitrogen oxide control), Stretford process for hydrogen sulphide removal (sulphur dioxide control); on-site solid waste disposal; on-site water treatment for recirculation to minimize discharge and external water requirements, and on-site shale oil upgrading facility. Oil shale facilities are subject to future regulations.

Resources used
(based on assumed 90 per cent capacity factor)

ANFO explosive	7.1×10^3 t/yr
Fuel for run-off oilshale mine	26.4×10^6 t/yr
Diesel fuel for on-site equipment	5.5×10^6 t/yr
Make-up water	44.7×10^9 l/yr
Land for mine development	4 ha) total over
Land for retorted shale disposal	400 ha) 25-year
Land for basic site facilities	56 ha) life span
Land affected by mining	2300 ha) of plant
Personnel required for construction	2300 people/yr
Personnel required for operation	1260 people/yr
Pollution control costs	US $0.9–1.7/bbl (syncrude equivalent)
Raw materials for construction)	
Facility costs)	not calculated

ENVIRONMENTAL RESIDUALS AND PRODUCTS

Air pollutants	Quantities released annually
SO_2	60 t
NO_x	1×10^9 t
Total suspended particulates (TSP)	51 t
Non-methane hydrocarbons (HC)	60 t
CO	358 t
Dust from shale storage	not determined
Particulates from retort	418 t
Carbon dioxide from retort	3.7×10^6 t

Table 6.7 (continued)

Post-control flue gas constituents	
$\cdot SO_2$	2.6×10^3 t
$\cdot NO_x$	1.9×10^3 t
$\cdot HC$	0.9×10^3 t
$\cdot CO$	1.6×10^3 t
$\cdot CO_2$	1.0×10^6 t
Water pollutants	
Run-off from shale storage	negligible
Run-off from retorted shale	433×10^6 l
Airborne water from cooling tower drift	
and evaporation	1.68×10^6 l
Total suspended solids	not determined
BOD	not determined
Oil and grease	not determined
Solid waste	
Retorted shale	20.9×10^6 tonnes
Thermal discharge	not determined
Noise pollution	not determined
Syncrude	2.5×10^9 litres
OCCUPATIONAL HEALTH AND SAFETY	
Risk of death	1.2/yr
Risk of injury	69.4/yr

(Source: US Department of Energy, 1988)

- size – 50×10^3 barrels/day (syncrude equivalent);
- annual capacity factor – 90 per cent (assumed);
- annual energy production – 106.4×10^{12} Btu;
- efficiency – 95 per cent (extraction);
- life span – 25 years.

Wood processing

The primary method of processing wood for fuel purposes is through carbonization, more commonly known as charcoal making. When wood is carbonized to produce charcoal, approximately 60 to 90 per cent of the energy that is contained in the wood is lost (MacDonald, 1989; Overend, 1986; ILO, 1985; Boutette and Karch, 1984; Earl, 1984). Charcoal is more easily transported and stored than fuelwood and burns with little or no smoke. These properties make it an attractive fuel for domestic use and also mean that it will be more widely used in urban centres.

To manufacture charcoal, wood is placed in a kiln or buried in the ground and covered to achieve close to anaerobic conditions. The wood inside the kiln is lit, giving off energy and producing by-products such as acetic acid, wood oil, gas, creosote, pitch and tar in addition to charcoal. The methane gas produced during carbonization is generally allowed to escape into the atmosphere along with any steam which is formed.

Large-scale charcoal making

Large-scale manufacture of charcoal can seriously threaten the sustainability of surrounding woodlands, as large areas may be clear cut to provide wood for commercial operations. This can lead to long-term problems with wood supply and increase soil erosion due to a loss of ground cover. Charcoal making also leaves behind fines and by-products such as tar and wood oil which can contaminate the soil. These problems are much greater when large-scale manufacturing processes are used. Small-scale charcoal production is greatly preferred from an environmental perspective, in addition to increasing the requirements for rural labour.

Mobility of carbonization technology

The mobility of technology used here refers to the ease of locating a technology where adequate wood supplies exist. Stationary technologies (e.g. beehive brick kiln, Missouri cement kiln, industrial retort) will require larger and larger areas of cutting as forested lands close to the kiln become depleted. Non-stationary kilns (e.g. traditional pit kiln, traditional mound kiln, improved or modified earthen kilns, steel drum kiln, steel ring kilns (Mark V and FTP)) provide a more environmentally benign approach to charcoal making, as they may be constructed where there are suitable supplies of wood (MacDonald, 1989). These kilns are usually of a smaller scale than stationary kilns.

Environmental impacts of charcoal manufacture

Charcoal manufacture creates the greatest potential for environmental degradation through the overcutting of forests. This is particularly a problem if stationary large-scale technology is used. Soil immediately under and surrounding the kiln may also become contaminated with charcoal fines and the by-products of charcoal manufacture. Table 6.8 summarizes these problems and poses potential mitigation measures.

Table 6.8 *Environmental impacts of charcoal manufacture*

Potential impact	Environmental protection measure
Impacts to the physical environment	
■overcutting of forested land	■avoid use of large-scale charcoal manufacturing techniques (e.g. brick kiln, Missouri cement kiln, industrial retort); use wood from plantations for charcoal making wherever possible
■contamination of soil	■conduct charcoal making on non-arable land to reduce any impact on soil fertility; collect fines left behind after charcoal is removed from kiln, retrieve by-products (e.g. tar, acetic acid) and sell if market exists

Briquetting of biomass, municipal waste and mining waste

Biomass briquetting or pelletizing is conducted under much higher pressure (in excess of 150 MPa) than peat or lignite compaction. The high pressure alters the internal structure and, to some extent, the chemical nature of the biomass. The process is one of both densification and compaction. Various forms of biomass residue may be processed in this manner (e.g. sawdust or agro-residues) as well as household wastes. Screw presses and piston presses are the most widely used technologies for this type of briquetting. The former are continuous low-capacity machines which work by extruding material through a continuous screw; the latter operate as continuous ram presses at much higher capacities of up to 5 tonnes per hour. Both raise the temperature of the material to a level where internal binders are generated from the lignins which form the vegetable cell walls. These hold the compacted material, eliminating the need for an externally added binder.

Refuse-derived material is commonly processed in ring presses similar to those used in cattle feed and pelletizing ores. The resultant pellets are called refuse-derived fuel (RDF) pellets and have been used in power stations and large industrial boilers.

All biomass briquettes and RDF are comparatively expensive because of the high unit capital costs of the plant and associated energy costs. They are marginally economic in most circumstances and have found only limited application, mainly for sawdust and wood wastes.

Briquetting or pelletizing of waste products can have a positive effect on the environment from the perspective of the recovery and use of wastes which would otherwise be discarded. Some of these wastes may have alternative uses, albeit of low value, such as agricultural wastes being used for animal bedding. They may also be burned in individual households, although often at a low efficiency.

Treatment and processing of liquid wastes

Liquid wastes are created by a number of industrial processes and may be used as fuels. They most commonly originate from pulp and paper manufacture (black liquor, clarifier sludge) and municipal or industrial wastewater treatment (sewage sludge). These liquids can contain pollutant compounds such as heavy metals that are released from the fuel in the combustion stage.

Processing of black liquor and clarifier sludges that are to be used as fuel involves evaporation in settling ponds to reach 50 per cent of the original moisture (Kiørboe, 1990). Sewage sludge is treated with chemical additives and polymer conditioning and water removed through mechanical means such as screw presses or belt filter dewatering (Kalmari, Maskuniitty and Kosunen, 1990; Walsh, Pincince and Niesseri, 1990). Following drying, the sludge may be used in a pulverized or cake form for combustion.

References

AARNA, A (1978) *Chemical Engineering in the Estonian SSR.* Perioodika, Tallinn.

ATTASSI, M (1992) 'Environmental impacts of the exploitation of El-Lajjun oil shales in Syria'. In *The Environmentally Sound Management of Low-grade Fuels.* SEI, Stockholm.

BCRA (1987) 'Beneficiation of low-rank coals'. *BCRA Quarterly* **15**: 41–66.

BORD NA MÓNA (1985) *Fuel Peat in Developing Countries.* World Bank, Washington.

BORON, D J, EVANS, E W and PETERSON, J M (1987) 'An overview of peat research, utilization and environmental considerations'. *International Journal of Coal Geology* **8**: 1–31.

BOUTETTE, M and KARCH, G E (1984) *Charcoal: Small-scale Production and Use.* Friedr Vieweg, Braunschweig.

CARTER, S D, GRAHAM, U M, RUBEL, A M and ROBL, T L (1995) 'Fluidized bed retorting of oil shale'. In *Composition, Geochemistry and Conversion of Oil Shales.* Kluwer Academic Publishers, Dordrecht.

COUCH, G R (1990) *Lignite Upgrading*. IEA Coal Research, London.

EARL, D E (1984) *Report on Charcoal Production*. Sudan Renewable Energy Project, Khartoum.

EKINCI, E (1995) 'Economic considerations of oil shale and related conversion processed'. In *Composition, Geochemistry and Conversion of Oil Shales*. Kluwer Academic Publishers, Dordrecht.

EPA (1981) *Environmental Assessment: Perspective on the Emerging Oil Shale Industry*. Environmental Protection Agency, Washington.

ERG, K (1994) 'Chemical composition of groundwater in the Estonian oil shale basin'. *National Conference Publications, Institute of Engineering, Australia* **2**: 657–659.

ERIKSSON, S and PRIOR, M (1990) *The Briquetting of Agricultural Wastes for Fuel*. FAO, Rome.

FOHL, J and TESSMER, G (1985) 'Drying brown coal in a saturated steam atmosphere: system Fleissner'. In *Coaltech 85*. Industrial Presentations, Shiedam.

GROPPO, J G (1995) 'Oil shale beneficiation for processing'. In *Composition, Geochemistry and Conversion of Oil Shales*. Kluwer Academic Publishers, Dordrecht.

HAMILTON, L D (1990) *Personal Communication*. Brookhaven National Laboratory, Upton.

HARMARNEH, Y (1985) 'Direct combusion of El-Lajjun oil shale'. *The Syrian Journal of Geology* **9**: 85–88.

HOUSE, P W (1983) *Report EP-0093*. US Department of Energy, Washington.

HOUSE, P W (1981) *Energy Technology and the Environment*. US Department of Energy, Washington.

HUETTENHAIN, H (1992) 'Low-rank coal upgrading technology review'. *Conference on Clean and Efficient Use of Coal: A New Era for Low-rank Coal*. IEA/OECD, Paris.

ILO (1985) *Fuelwood and Charcoal Preparation*. International Labour Organization, Geneva.

IWG CORPORATION (1982) *Health and Environmental Effects Documents on Oil Shale*. IWG, San Diego.

KALMARI, A, MASKUNIITTY, H and KOSUNEN, P (1990) 'Handling systems for peat, woodwaste and sludge in industrial projects'. *Low-grade Fuels*. Vol 1. Technical Research Centre of Finland, Espoo.

KIØRBOE, K G (1990) 'Danish experiences with combustion of low-grade fuels'. *Low-grade Fuels*. Vol 1. Technical Research Centre of Finland, Espoo.

LEE, S (1991) *Oil Shale Technology*. CRC Press, Florida.

LURGI (1988) *Oil Shale Retorting: The Lurgi-Ruhrgas (LR) Process*. Lurgi GmbH, Frankfurt.

MACDONALD, M E (1989) *An Assessment of Higher Yield Charcoal Manufacturing Technology in the Sudan*. IES/Ecoville, Toronto.

OVEREND, R (1986) 'Bioenergy conversion process: a brief state of the art and discussion of environmental implications'. *Proceedings of International Union of Forestry Research Organization*. Ljubljana, Yugoslavia.

PENNER, S S (1982) 'Assessment of research needs for oil recovery from heavy oil sources and tar sands'. *Energy* **7**: 567.

PIERCE, N (1993) 'Waste management challenges in Russia, Ukraine and Estonia'. *Waste Age* **24**: 194–198.

PREIS, S (1994) 'Oxidative purification of wastewaters containing phenolic compounds from oil shale treatment'. *Environmental Technology* **15**: 135–144.

PROBSTEIN, R F and HICKS, R E (1982) *Synthetic Fuels*. McGraw Hill, New York.

PURRE, T A and VASILYEV, V V (1992) 'Baltic oil shale semicoke tar processing'. In *The Environmentally Sound Management of Low-grade Fuels*. SEI, Stockholm.

ROBERTSON, R A and GODSMAN, N M (1975) *Peat as an Energy Source in Scotland*. Macaulay Institute for Soil Research, Aberdeen.

SONDREAL, E A (1992) Clean utilization of low-rank coals for low-cost power generation. *Conference on Clean and Efficient Use of Coal: The New Era for Low-rank Coal*. IEA/OECD, Paris.

TSAROS, C L (1981) 'Peat dewatering: an overview'. *Proceedings of Symposium on Peat as an Energy Alternative*. Arlington, Virginia.

UNEP (1985) *The Environmental Impacts of Exploitation of Oil Shales and Tar Sands*. UNEP, Nairobi.

US DEPARTMENT OF ENERGY (1988) *Energy Technologies and the Environment: Environmental Information Handbook*. US Department of Energy, Washington.

WALSH, M J, PINCINCE, A B and NIESSERI, W R (1990) *Fuel-efficient Sewage Sludge Incineration*. EPA, Cincinnati.

YEN, T F (1976) *Science and Technology of Oil Shale*. Ann Arbor Science Publishers, Ann Arbor.

YEN, T F and CHILINGARIAN, G V (eds) (1976) *Oil Shale*. Elsevier, Amsterdam.

ZHIRYAKOV, Y N, RYATSEP, A Y and AUNAP, A K (1992) 'Ecological aspects of oil shale processing'. In *The Environmentally Sound Management of Low-grade Fuels*. SEI, Stockholm.

7

Environmental management and low-grade fuel conversion

The primary form of conversion for the solid low-grade fuels discussed here is combustion. Low-grade fuels may also be transformed from a solid to a liquid or gaseous state through a variety of techniques which are usually designed to raise the calorific value of the fuel. The potential major impacts associated with fuel conversion include a decrease in air quality due to the formation and release of emissions carrying impurities such as SO_x, NO_x, particulates, heavy metals and PAHs and the build-up of solid wastes originating from the mineral matter of the fuel. Great amounts of solid waste can require large quantities of land for disposal and may also contain toxins and pollutants such as heavy metals. In addition, the negative impacts on public health are greatest at this stage of the fuel cycle.

A wide variation in the composition of low-grade fuels between and within sources both chemically and physically means that it is particularly difficult to burn these fuels in an efficient way. Inefficient fuel combustion is not only wasteful but leads to higher levels of pollutant emissions, particularly those compounds that are a result of incomplete oxidation such as PAHs, some of which are known carcinogens and mutagens. The information which follows provides details of methods of low-grade fuel conversion. The major pollutants and solid wastes created during fuel conversion must be controlled to protect both public health and the well-being of the potentially affected ecosystems. Therefore, suggestions for environmental protection at this stage of the fuel cycle such as pollutant control technologies are also given.

Low-grade fuel conversion technologies

Low-grade fuels are currently in use in many parts of the world to produce energy to meet domestic, industrial and power production needs (e.g. Petzel, 1994; Chiewwattakee, 1993; Hajicek et al., 1993; Ji, 1993; Karekezi *et al.*, 1991; Kiørboe, 1990; Leach and Mearns, 1988). The technologies described here are those most frequently mentioned in the literature over the last decade.

The vast majority of power-generating systems using solid fuel employ pulverized fuel and fixed bed or grate combustion. The variability in the composition of low-grade fuels such as lignite, peat and oil shale has historically caused operational problems with conventional boiler systems due to slagging and fouling and erosion of components. Newer technologies such as fluidized bed combustion (FBC) or integrated gasification combined cycle (IGCC) have been developed that can adjust to variations in fuel quality and reduce the need for fuel pulverization and flue gas scrubbers. More fluidized bed combustors and pressurized fluidized bed combustors are coming on line; however, it has been estimated that only 3 per cent of current power generation takes place in FBC systems (Couch, 1992).

Low-grade fuels have been used in all of the conversion technologies discussed below, with varying degrees of success. PFBC and IGCC are complex and expensive systems which continue to undergo refinement. Although there is a general feeling of optimism regarding the use of these systems to reduce and eliminate some but not all pollutant emissions and effluents as well as increase boiler efficiency, their use is presently limited to several test plants.

Lignite has been combusted in all the technologies available for coal and is used for power production in many parts of the world. Oil shales have not been widely exploited to produce energy, and use has been primarily limited to Estonia and the Great Western Plains region of the United States. The high ash content of oil shale (up to 50 per cent) and the presence of toxic substances in the ash make oil shale difficult to burn. The most effective approach for power generation has been the use of a fixed bed boiler with a tower design where heating surfaces are arranged so that fouling problems are reduced (Ots, 1992).

Peat has been fired in grate, cyclone, special boilers for pulverized fuel and fluidized bed boilers with the most common option being pulverized fuel systems. Peat boilers are larger than conventional coal-fired boilers, since a longer time is required in the boiler to facilitate complete combustion and to reduce problems of fouling of the heating surfaces (EPRI, 1983).

Boilers at wood-fired district heating plants are usually in the range of 1–20 megawatts (thermal) (MW(th)) and consist of grate combustors fed by screw stokers or spreaders. The combustion of MSW is carried out for district heating. MSW is also burnt for co-generation in a number of cities in Europe and Scandinavia, primarily using conventional fixed bed boiler systems. Biomass such as straw is converted into energy for district heating using a grate boiler, for example, in Denmark. In Sweden, Finland, the United States, Canada and the former Soviet Union the pulp and paper industry has burnt up to 100–150 tonnes per hour of bark in both fixed and fluidized bed combustors to produce steam for the co-generation of electricity.

Various forms of IGCC have been tested on all low-grade fuels. The largest project of this type, to date, is the Great Plains Coal Gasification project in North Dakota, where lignite is gasified to produce SNG. IGCC using circulating fluidized bed gasification has shown the most promise with wood, biomass and waste.

All fuel conversion processes require special design features such as flue gas treatment to ensure environmental protection from emissions, effluents and by-products created during their use. These are discussed above.

Stoves for domestic use of low-grade fuels

Wood, peat, biomass and lignite (usually in the form of briquettes) are widely converted in small stoves or fireplaces to provide energy for cooking food and, to a lesser extent, for space heating. Cooking food may not only improve its taste but also provides health benefits, since heating food to an appropriate temperature can kill bacteria and parasites which may be present in uncooked food. It is uncertain what percentage of low-grade fuels are converted in small cookstoves, but the impact of this activity on the quality of daily life, particularly where energy generated from power stations is unavailable or expensive, can be enormous.

The household method for converting low-grade fuels can range from a simple wood fire constructed on the bare ground to heavy iron stoves with high-precision mechanisms for controlling oxygen flow and heat output.

Fixed bed combustion

Traditional approaches to the combustion of solid fuel rely on the principle of supplying heat and air to fuel which is sitting on an open grid or grate. The size of the holes in the grid need to be small enough to hold the fuel and large enough to allow ash to fall out as the fuel burns. Feeder systems for low-grade fuel-fired plants must be designed to accommodate the large feed requirements that the moisture and ash content of the fuel necessitate (Johns, Clocker and Levslek, 1992, 1984; Generator Industri, 1985a, 1985b). The grate may sometimes shift to shake down the ash. In some places fuel may still be fed into the combustion system by hand, although most units now make use of mechanized stoking. The grate may also move continuously in a line in such a way that it 'tilts and doubles back on itself' (Patterson, 1987). This type of combustion is known as a moving chain grate stoker (see Figure 7.1).

Pulverized fuel The size of the individual fragments of fuel which undergo combustion has an impact on combustion efficiency, since a larger surface area presents greater opportunities for reactions to occur. Breaking fuels down into smaller fragments, even to the point of fine powder, creates a

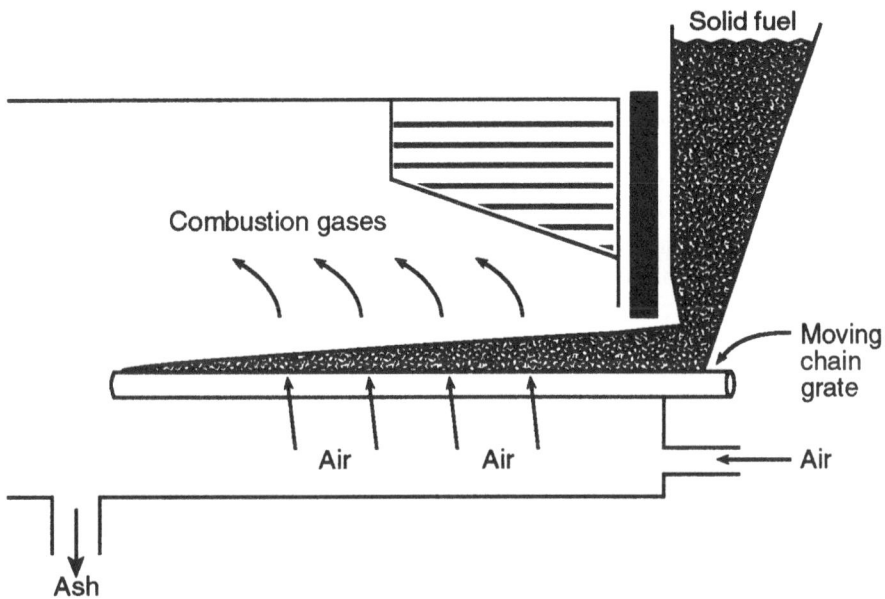

Figure 7.1 *Chain grate stoker*

(Source: Patterson, 1987)

greater surface area for reactions and is known as pulverizing fuel. Many conventional combustion units make use of this treatment of the fuel to increase the amount of heat released during combustion of a given unit of fuel (by weight).

Fluidized bed combustion

During the early 1960s the burning of low-grade fuels on a fluidized bed began to attract attention due to the efficiency of fuel conversion connected with this process and the ability of the technology to accept fuels with variable characteristics such as increasingly high ash levels (Boyd, 1994; Hajicek et al., 1993). Over time it became apparent that fluidized bed combustion had the added advantage of reducing emissions of SO_2 and NO_x when burning low-grade fuels (Riedle and Böhm, 1992; Geisler, Heindrich and Gonderman, 1990; Poersch and Zabescheck, 1980).

FBC systems have traditionally been categorized on the basis of the velocity of the air entering the combustion chamber and the impact of the moving air on the fuel bed. In bubbling fluidized bed combustors the air enters the combustion chamber relatively slowly, causing the fuel to shift and bubble. Circulating fluidized bed combustion (CFBC) is defined by rapidly moving air which enters the combustion chamber with such a force that the fuel particles remain suspended in the chamber. Although both types of FBC are currently in use, more recent innovations have combined elements of each approach, thus reducing the distinctions between the two (Makansi, 1991).

In a typical bubbling FBC process the fuel, which can be solid, liquid or gaseous, is kept suspended during combustion through the presence of air originating from below the combustion floor with a velocity depending upon design of up to 1.5–2 m/sec. The suspension of the fuel results in a high rate of mixing between the fuel and the oxidant. At a high mixing rate heat is generated at a substantially lower, and more uniform, temperature than in a fixed bed-fired unit.

The low temperature of the combustion bed (850–930°C) reduces the amount of slagging or fouling of the bed and inhibits formation of thermal NO_x. In addition, bubbling FBC does not require prior treatment of the low-grade fuel (e.g. pre-drying, milling or blast pre-heating) nor is it necessary to use a supporting fuel to ignite or stabilize the flame. An FBC system is illustrated in Figure 7.2.

Circulating fluidized bed combustion

Circulating fluidized bed combustion is also known as fast fluidized bed combustion, since the velocity of the fluidizing gas rushing into the combustion chamber is much higher than for bubbling FBC. The velocity of the gas ranges from 5 to 10 m/sec (Howard, 1989) to 15 m/sec (Patterson, 1987) and, in the former Soviet Union, sometimes reaches levels of up to 25 m/sec.

A high fluidizing gas velocity promotes the carrying of particles from the bed of combustion high into the combustion chamber (a process known as elutriation). This leads to the fuel having a longer residence time in the combustion chamber, which increases the opportunity for efficient burning of carbon and, if an adsorbent such as limestone or dolomite is added to the chamber, excellent recovery of SO_2 may occur. CFBC also allows for staged combustion, which generally aids in suppressing NO_x formation though not N_2O, since staging permits low oxygen levels during each stage.

Experience over the last decade from major CFBC producers such as Ahlström has indicated that CFBC effectively converts many types of fuel, including wood, peat, oil shale, lignite, agricultural wastes and MSW into heat, with low levels of pollutant emissions and a relatively small build-up of solid waste (Darling, 1994; Murphy, 1994; Coulthard, Korenberg and Oswald, 1991; Ahlström, 1990; Eipper, 1990, 1986; Manzoori, Linder and

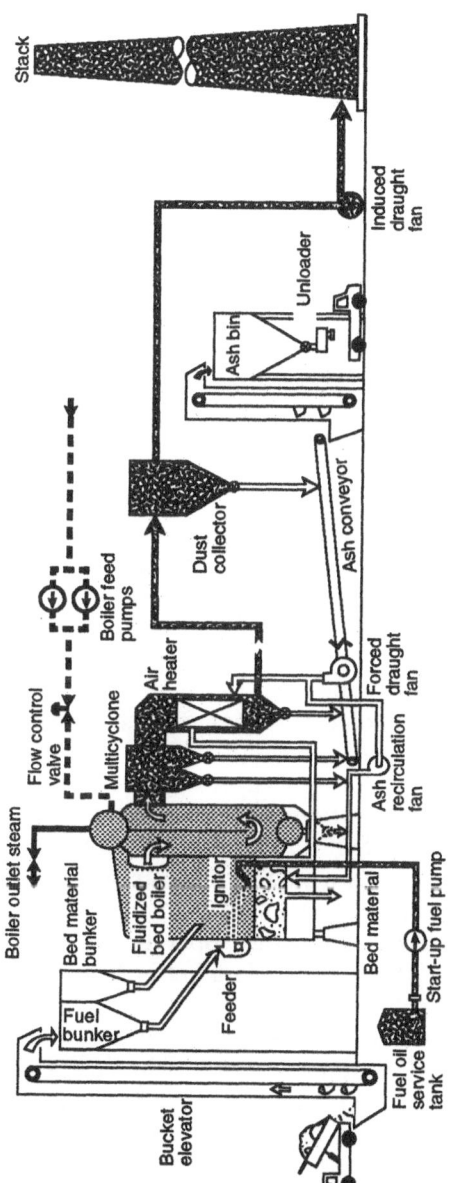

Figure 7.2 *FBC system*

(Source: Ishikawajima-Harima Ltd., 1991)

Agarwal, 1989). Users of CFBC are encouraged to follow the manufacturer's guidelines as closely as possible and to continually review performance requirements to ensure maximum combustion efficiency and environmental protection (Seitzinger and Morrison, 1993).

Design features An example of a CFBC system is given in Figure 7.3. Four basic components can generally be distinguished in CFBC technologies, which are described below.

Combustor vessel The combustor vessel is divided into two zones – a lower reducing zone and an upper oxidizing zone. In the lower reducing zone primary combustion air is introduced by an air distribution grid. The quantity of air used at this point is less than that required for stoichiometric combustion of the fuel. The lower reducing zone serves to minimize fuel NO_x and thermal NO_x generation and reduces fan power consumption. In the upper oxidizing zone the air required to achieve complete combustion is introduced through suitable nozzles in the vessel wall into the suspended gas, fuel and air.

Cyclone There are several types and positionings of cyclones within different CFBC designs. Cyclones are used to remove unburned fuel, particulates and charcoal prior to release of the flue gases. High-efficiency hot cyclones are refractory lined and allow passage of the hot circulating solids to the external exchanger. The hot flue gases leave the top of the cyclones and escape through a duct to the conventional convective boiler system to generate steam. Water-cooled cyclones may also be incorporated into the boiler design.

External heat exchanger (optional) In CFBC the tubes of the heat-exchange surfaces are located outside the combustion chamber to reduce any erosion of the heat-transfer surfaces which may be caused by fast-moving particles from the circulating bed. External heat exchanges allow for the recovery of heat from the hot flowing ash and separate the heat transfer from the combustion process, which results in a better heat end load.

Convective heat transfer surface, superheater, air pre-heater These devices are built into the CFBC system to regulate the heat within the process in a manner which ensures high combustion efficiency either through heat absorption or air heating. The superheater may be situated in the combustion chamber ahead of the cyclone, an area where heat transfer rates are high. This configuration will reduce the temperature of the flue gas going into the

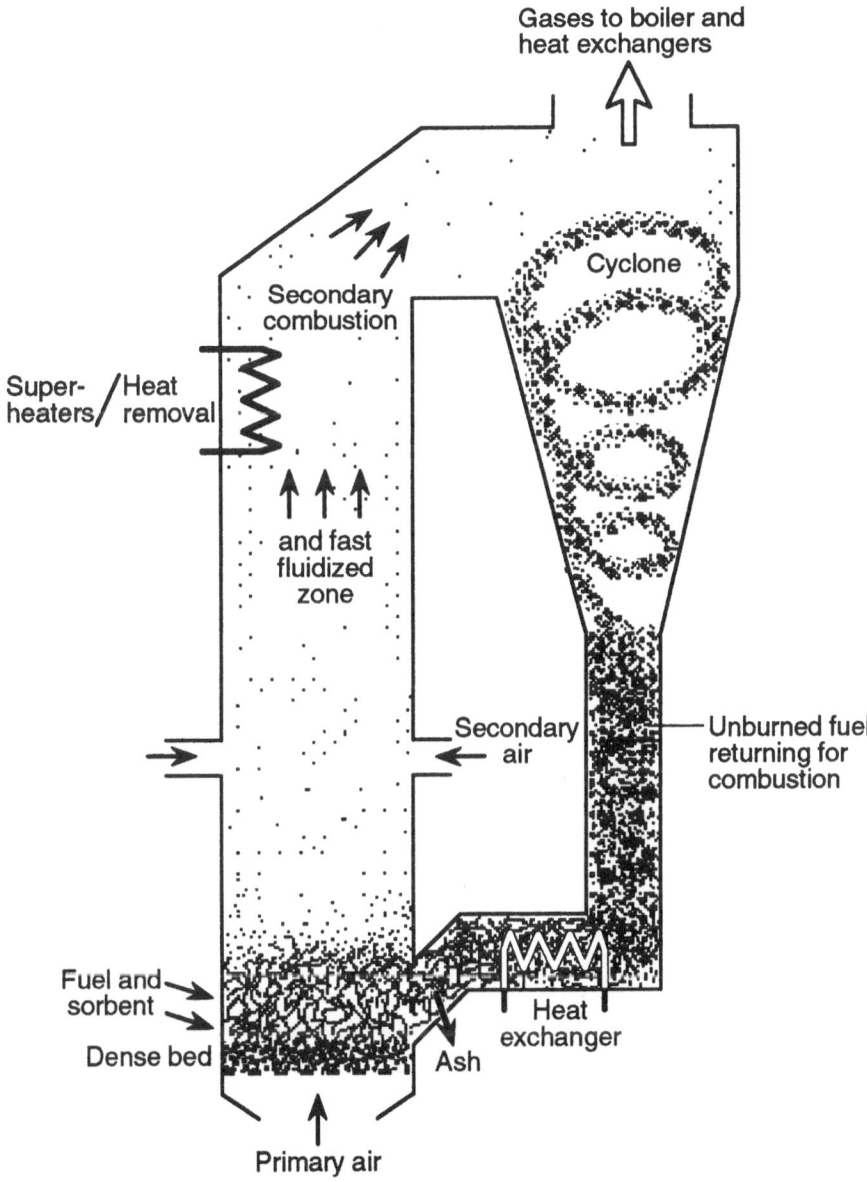

Figure 7.3 *CFBC system*

cyclone and requires a smaller cyclone than high-temperature flue gases (Makansi, 1991).

Air pre-heating may be accomplished through tubular air heaters, plate-type air heaters or heat-pipe-type air heaters. The pre-heating of air may take place during different stages of the CFBC system, depending upon the individual design and requirements of the operation.

Pressurized fluidized bed combustion

In a PFBC system fuel is burned under pressure in a fluidized or circulating bed. This allows combustion to occur at a more intense level per unit volume compared with FBC or CFBC. Pressurized fluidized bed units are smaller than other forms of combustion per unit of energy produced, but they do present engineering problems with the feeding of fuel and the removal of ash through pressure locks into and out of the combustion chamber (Patterson, 1990).

PFBC is an expensive technology which is not yet fully proven (Branwell, 1991; Patterson, 1990), particularly with respect to low-grade fuels. Like most solid fuel combustion technology, tests for PFBC were conducted on higher-grade coals such as bituminous coals, whose properties differ significantly from lignite, peat, oil shale or wood.

The most extensive use of a low-grade fuel in PFBC has been in Spain, where lignite with an ash content of 25–45 per cent and a sulphur content in the range of 4–8 per cent is being used to fuel an 80 MW retrofit system. The properties inherent in this lignite necessitate the feeding of lignite and limestone into the combustion chamber in dry form rather than in the traditional wet paste form employed with other coals (Jansson, 1992, 1990; Patterson, 1990).

Low-grade fuel gasification

Most low-grade fuels may be converted to a gas to meet a variety of end uses. Gas is a difficult, space-consuming and often expensive material to transport and store. For domestic purposes and small-scale industry biogasification is the most common gasification practice. For larger industries or power production, gas manufactured from low-grade fuels is generally most cost-effective if production of the gas takes place as part of the process of energy production. This approach to gasification is known as combined cycle power generation or integrated gasification combined cycle.

Biogasification The manufacture of biogas involves the fermentation of organic materials under anaerobic conditions to produce methane, which can be burned. In many developing countries, and in India and China in particular, biogasification is used to derive energy from materials such as crop wastes, rice husks, nut husks, coffee grinds and dried plants such as straw or water lilies. Digestors for biogas production were introduced and encouraged in a number of developing countries by several international aid agencies in the late seventies and early eighties. Many of these digestors have had technical problems over time, although the reasons behind these problems are not well understood (IDRC, 1986; Overend, 1986).

A major drawback to biogasification involves the use of dung to produce biogas instead of being applied to the soil to enhance soil quality and contribute to increased crop production. Changing the amount of dung added to the soil could have long- and short-term effects on soil fertility. In larger biogasification programmes plant material for gasification may be grown on land which would otherwise be used to produce food. As Figure 7.4 illustrates, it is important to confine biomass production for fuel to non-arable land, particularly where arable land is a scarce resource.

Integrated gasification combined cycle

Low-grade fuels may be transformed to a gaseous state for use by industry or for electricity production. The gasification of low-grade fuels as part of a larger system of energy production is known as integrated gasification combined cycle (IGCC). IGCC has been carried out, either in tests or on a commercial basis, using lignite, peat, biomass, municipal waste, RDF and briquettes made from sub-bituminous coal fines. Figure 7.5 illustrates the conceptual scheme of an IGCC. There are many variations on this theme, particularly with respect to the choice of gasifier.

The primary benefits of using an IGCC include:

■ increased combustion efficiency;
■ low water requirements (less than 50 per cent of conventional approaches);
■ smaller land requirements; and
■ reduction of sulphur, NO_x and particulate emissions
(Beer and Homola, 1992; Patel and Mensinger, 1990).

Success with the IGCC using lignite as a fuel has been achieved with the Dow Syngas project and the Great Plains Coal Gasification project, both located in the United States.

Fixed bed gasification The fixed bed gasifier illustrated in Figure 7.6 is a double-walled, water-cooled reactor which makes use of steam and oxygen

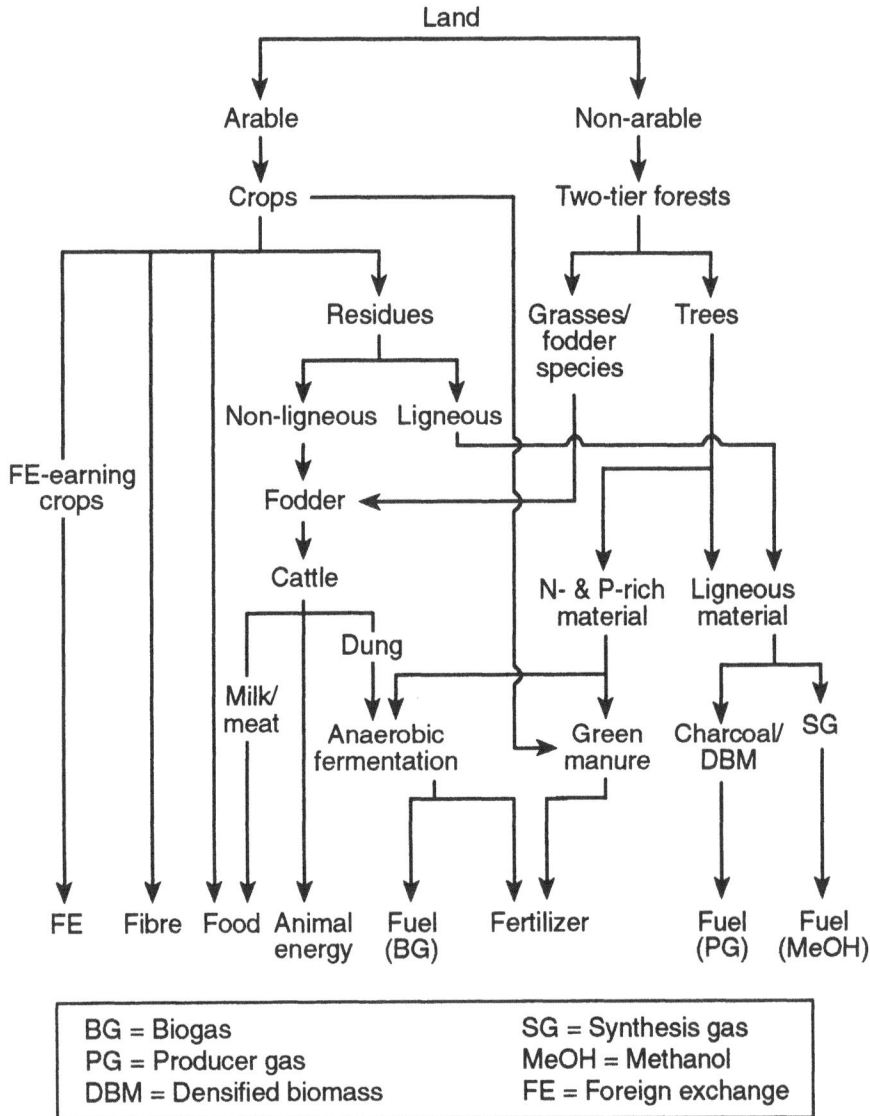

Figure 7.4 *A tentative set of decision rules for biomass production in countries with land constraints*

(Source: Reddy, 1985)

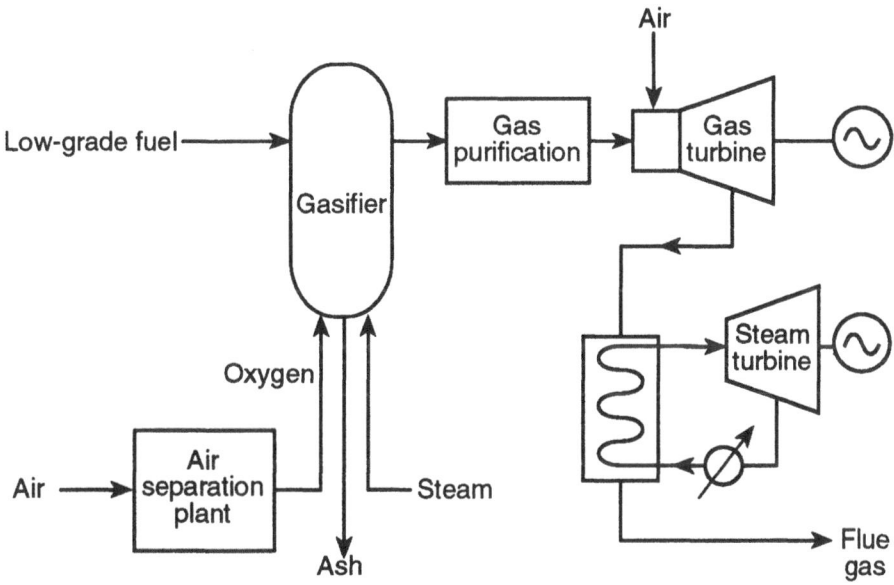

Figure 7.5 *IGCC system*

(Source: Patel and Mensinger, 1990)

to create conditions for fuel gasification. This method is known as the British Gas/Lurgi slagging gasifier and has been used to gasify fuel, up to 50 per cent of which is comprised of fines compressed into briquettes prior to gasification (Herbert, Loeffler and Wechster, 1988; Lurgi, 1988).

Circulating fluidized bed gasification Gasification of low-grade fuels may also be carried out using a CFB gasifier. Peat, wood waste, sawdust or bark can be used in a CFB gasifier. These units are compatible with industrial power production and are operating successfully at a number of pulp and paper mills in Sweden and Finland, where waste from pulp and paper production is suitable for energy production at the mill. Figure 7.7 illustrates a 34-MW CFB gasifier using residue from pulp and paper manufacture in place at a Swedish pulp and paper mill.

Pressurized fluidized bed gasification The use of a fluidized bed gasification system under pressure has proven to be an effective and efficient gasification process for use in an IGCC system. The majority of tests on this process, known as the High Temperature Winkler (HTW) process, have been conducted by Reinische Braunkohlenwerke AG using

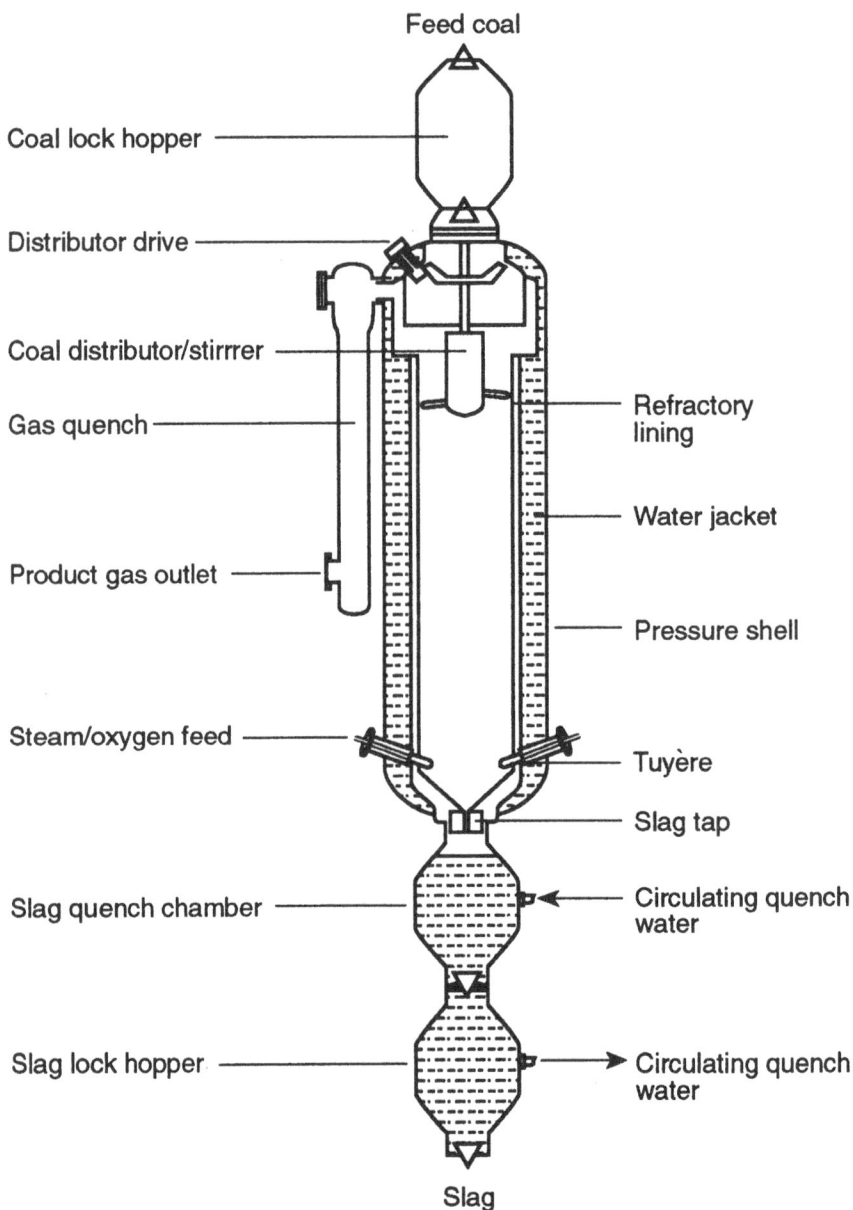

Figure 7.6 *British Gas/Lurgi slagging gasifier*

(Source: Patel and Mensinger, 1990)

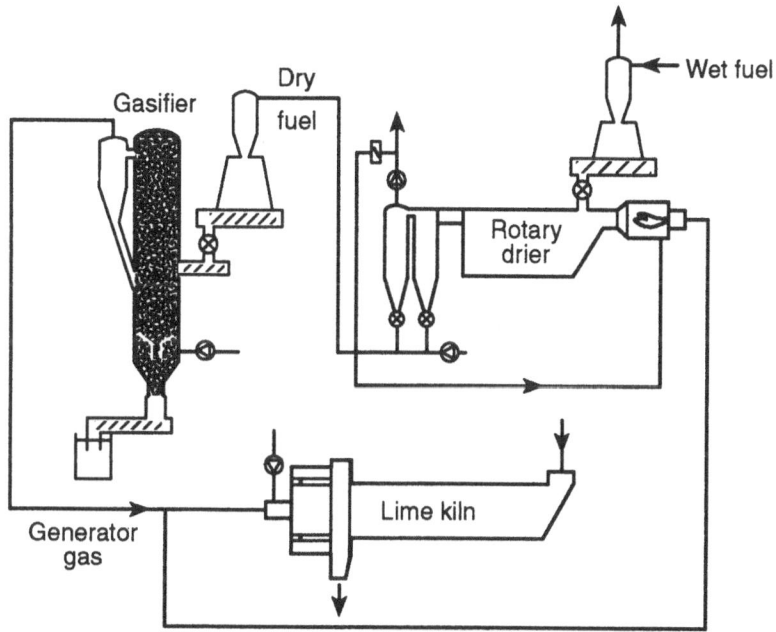

Figure 7.7 *IGCC with a CFB gasifier at a pulp and paper mill*

(Source: Götaverken Energy, 1991)

lignite (Rhenish brown coal) for power production. This approach to gasification has not yet reached the stage where it is commercially available (Aslhock, Keller and Herbert, 1990; UHDE, 1987). Table 7.1 outlines the fuels investigated for use in HTW gasification and the level of testing these fuels have received to date.

Fuel pyrolysis

The thermal destruction of the organic matter in solid fuels, under anaerobic conditions, is known as pyrolysis. During pyrolysis large quantities of volatile vapour and gaseous compounds are released. The temperature of pyrolysis has a great effect on the process, so that pyrolysis carried out at different temperatures is known by specific names. Types of pyrolysis include:

- bertination (350–450°C)
- low-temperature carbonization (450–650°C)

Table 7.1 *Feedstocks investigated regarding their suitability for HTW gasification*

Fuel	Country	Laboratory investigations	Tests in the PDU	Tests in the pilot plant	Tests in the demonstration plant	Study for combined cycle
Group I (brown coals)						
■Rhenish brown coal	Germany	X	X	X	X	X
Coke from Rhenish brown coal	Germany	X				
■Maramarua coal	New Zealand	X	X	X	X	
■Yunnan coal	China	X				
Group II (others)						
Wood	Brazil, Sweden	X	X			
■Wood	Kenya, USA	X				
■Peat	Finland	X	X	X	X	
Peat	USA, Canada	X				
■High sulphur- and salt-containing brown coal	South Australia	X	X	X		X

Table 7.1 (continued)

Fuel	Country	Laboratory Investigations	Tests in the PDU	Tests in the pilot plant	Tests in the demonstration plant	Study for combined cycle
High ash-containing brown coal	Greece, Australia, Spain	X				
Hard brown coal (lignite)	USA, Canada	X	X			
Hard brown coal (lignite)	China, Spain, New Zealand, Australia	X				
High volatile bituminous coal (sub-bituminous)	Australia, South Africa	X	X			
■Coking hard coal	Germany	X	X			
■Coking hard coal	China	X	X			
■Residue from coal/ oil hydrogenation	Germany	X	X			
■Sludge	Germany	X	X			

(Source: Theis and Lambertz, 1990)

- middle-temperature carbonization (650–750°C)
- carbonization (1100–1300°C)

The three main stages of pyrolysis are fuel heating, chemical destruction of organic material and some mineral matter in the fuel, and secondary processes such as chemical reactions in the vapour and gaseous products formed during pyrolysis. The temperatures at which different fuels begin to release volatiles is given in Table 7.2.

Table 7.2 *Initial temperature of devolatization during pyrolysis*

Fuel	Temperature (°C)
Peat	100–110
Biomass	160
Lignite	130–170
Oil shale	250
Anthracite	380–400

The volatiles released during pyrolysis include hydrocarbons (gaseous and condensed), CO and CO_2, hydrogen, hydrogen sulphide and water vapour. The quantity and composition of volatiles that escape during pyrolysis depend on the final heating temperature of the process, the duration of heating and the rate of product removal from the reaction zone. It is possible to influence, to a certain degree, the character and amount of volatiles released during pyrolysis by adjusting these parameters. Rapid heating (e.g. heating of peat or lignite to 400–500°C in several seconds) is the method of pyrolysis most conducive to the production of chemically valuable components. Under these conditions there is only a small loss of potentially useful hydrocarbons through evaporation.

Solid residue which is also formed during pyrolysis has a higher carbon and ash content than primary raw fuel. Depending upon the final temperature solid residue from pyrolysis may be called bertinate (350–400°C), char or semicoke (550–650°C), mid-temperature coke (700–750°C) and coke (greater than 900°C). Figure 7.8 outlines the product release material balance occurring during pyrolysis.

Pyrolysis may be combined with the gasification or combustion of solid residue. This happens most frequently with the use of multistage fluidized bed reactors where solid fuels are heated up to 315–670°C without access to air. The recirculation of hot char ensures the absorption of heat by the fuel bed.

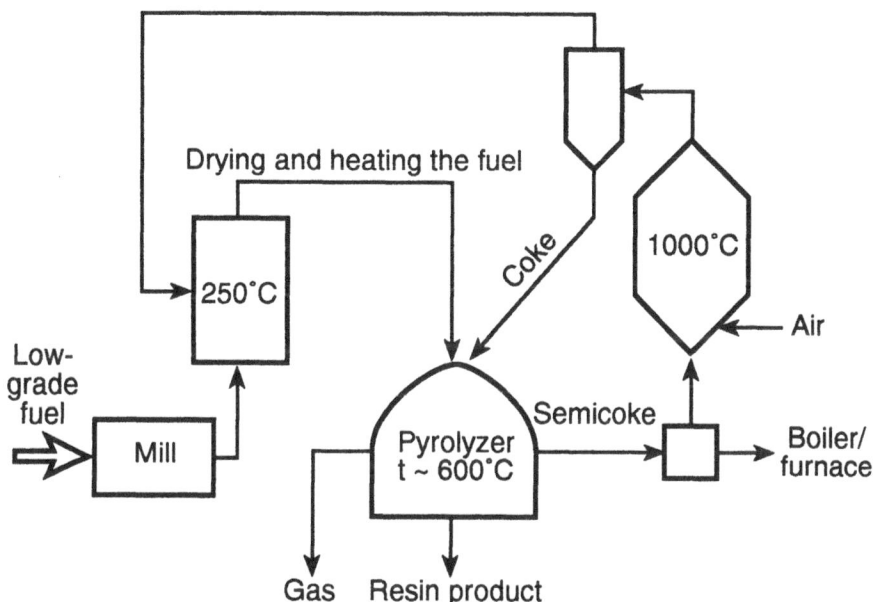

Figure 7.8 *Process and products of pyrolysis*

Although fuel pyrolysis is implemented in a number of countries as a separate technological process for chemical production, in the former Soviet Union the development of rapid pyrolysis technology for low-grade fuels has primarily focused on power production. Thermal efficiency is greatest when these plants are operated on a large scale, since high amounts of solid by-products which are produced by the process may be used to fuel the power plant.

Hydropyrolysis Pyrolysis can also be used to create liquid products through the use of hydropyrolysis. With hydropyrolysis solid fuel undergoes chemical destruction at 480–490°C under hydrogen pressure and, in some cases, in the presence of a catalyst. Most of the organic material in the fuel is converted to liquid hydrocarbons in this manner. At 6 per cent hydrogen the liquid yield will usually be 55–60 per cent per unit of organic matter. This liquid would typically comprise 10–15 per cent phenols, 3–5 per cent nitrogen compounds and 30–50 per cent aromatic hydrocarbons, which may be subjected to hydrogenation to produce automobile or aviation fuel.

An overview of pollutant formation and control

There are a number of ways in which low-grade fuel conversion products can pollute air, water and soil in the environment adjacent to a conversion plant or be carried by water or wind to more distant locales. These pollutants can cause health and environmental impacts on a global, regional or local level. The potential for the formation of pollutant substances during combustion is dependent upon the composition of a low-grade fuel and on the processes used to release the energy contained in the fuel. The following sections present a discussion on specific pollutants formed and/or released during combustion, including particulate matter, CO_2, SO_x, halogens, volatile organic compounds, gaseous trace elements and solid wastes. An example of the potential impacts to air quality from low-grade fuel combustion is summarized in Table 7.3.

Carbon dioxide, a product of fuel combustion, and halogens such as chlorine and fluorine, which may be released during combustion, make a significant contribution to the build-up of greenhouse gases in the atmosphere. Pollutants such as SO_2 and NO_x may have a detrimental effect on plant communities, thus causing a disruption to fish and wildlife habitat. The production of N_2O has been linked to ozone depletion. The potential also exists for an impact on human health through the intake of heavy metals such as lead, which may remain, or mercury and cadmium, which may be released in volatile form following the burning of a low-grade fuel. PAHs, many of which are known carcinogens, may form during incomplete combustion and will pose a health hazard if they are inhaled or come in contact with the skin.

In addition to the pre-combustion cleaning or upgrading of low-grade fuels described in Chapter 6, there are several basic approaches to preventing, controlling or eliminating pollutants during fuel conversion. Overall requirements for efficient combustion and effective control must take into account the '3-T rule', that is, time (residence time for fuel/gas mixture in active combustion zone), turbulence (mixture of air and burning gases) and temperature (adequate burning temperature).

Options for pollution control include:

- retention of pollutants by direct injection of absorbents during the combustion stage;
- removal of mineral matter and fuel heterotoms through intra-cycle fuel processing; and
- cleaning, containing or removal of undesirable substances from combustion products at the 'back-end' of the cycle.

Table 7.3 *Potential impacts on air quality of low-grade fuel combustion*

Potential impact	Environmental protection measure
Decrease in air quality	
■ increased emissions of CO_2	■ commitment to conservation of fuel and high end-use efficiency
■ increased emissions of SO_2	■ use physical containment such as fabric filters (baghouses) or electrostatic precipitators to collect sulphur particles; flue gas desulphurization through dry methods; dry alkali injection into the combustion chamber; the addition of alkali chemicals dissolved in water (also known as wet scrubbing)
■ increased emissions of NO_x	■ suppression of NO_x formation during combustion by lowering combustion temperatures or controlling available oxygen levels; formation of NO_x from nitrogen in fuel may be controlled by flue gas treatment using methods such as catalytic and non-catalytic reduction
■ increased emissions of halogens	■ uncertain regarding control mechanisms, but current flue gas desulphurization using lime or limestone aids in reduction of halogen emissions
■ increased emissions of PAHs	■ at high combustion temperatures these combine with oxygen and become inert
■ increased emissions of heavy metals such as lead, cadmium and nickel	■ use of particulate control devices such as baghouses or electrostatic precipitators to capture trace elements; evidence suggests that flue gas desulphurization also reduces emission of heavy metals

When selecting technologies for controlling or eliminating emissions and effluents from a conversion plant, it is important to recognize that no two situations are identical. Site-specific considerations include:

■   fuel characteristics;
■   age of system;

- available technologies;
- boiler configuration;
- available space;
- standards and regulations dictated by current legislation and international protocols;
- whether plans are for retrofitting or upgrading an existing plant or building a new one;
- size of budget.

Of all the parameters listed above, overall environmental requirements related to the fuel composition may be the prime determinant of control technology. For example, selective catalytic reduction is effective in reducing NO_x when low-sulphur coals are used but not as effective with high-sulphur coals of the United States (EPRI, 1991).

With respect to plant upgrading the prevailing wisdom is to 'do the simple things first' (Couch, 1992). If high-cost FGD is not affordable, it is better to install an electrostatic precipitator or spray dryer than nothing at all, even though these options may be less efficient than the most up-to-date technology. Upgrading plans will frequently require a discussion between the governing authorities and the plant owners to reach agreement regarding the most appropriate solution.

Pollution control equipment can also produce secondary pollution in the form of ash, residues and sludges. Some of these substances will require treatment, while others may be offered for sale (e.g. gypsum or sulphuric acid). A consideration of solid waste produced as a result of low-grade fuel conversion is given below.

Health risks of low-grade fuel combustion

The greatest risk to health from low-grade fuel use is associated with the combustion stage of the fuel cycle (IAEA, 1991). Information for coal, oil shale and peat indicates that risks to health are present for both those working with the fuels and the general public. The primary risk for workers at power plants is loss of hearing and accidents while operating machinery. At lignite power stations lignite dust and asbestos place plant workers at risk of respiratory irritation and diseases.

For all low-grade fuel use, ranging from wood used in a domestic setting to large-scale power production, the principal risk to public health arises from exposure to pollutants formed during combustion which may contribute to respiratory diseases and cancer. For peat, emissions may include particulate matter and benzo(a)pyrene (Hamilton, 1990; UNEP, 1985a; Lautkaski et al., 1982). Lignite combustion may enable release of SO_2, CO, NO_x, hydrocarbons, polycyclic aromatic matter, fly ash, trace metals and radionuclides (UNEP, 1985a; Hamilton, 1984; Hubert et al., 1981; Black and Neihaus, 1980; Morris, 1979). For oil shale, potential pollutants include carcinogenic heavy metals (e.g. arsenic, cadmium, chromium and nickel), PAHs and radionuclides (Hamilton, 1990; UNEP, 1985b; House, 1983, 1981; IWG, 1982). An additional health risk to the public may occur from the contamination of drinking water from spent shale storage piles. The high heavy metal content of municipal waste and industrial sewage sludge, often the highest of all low-grade fuels (Häni, 1991; Vancil, Parrish and Palazzolo, 1991), points to the likelihood of a risk to health arising from the burning of these fuels.

Table 7.4 indicates the health risks associated with the combustion of coal, peat and oil shale. From these figures it appears that there is less risk for workers and the public from the combustion of peat than from either lignite or oil shale. The table also shows that there is a wider range of risk for workers and the public from coal and oil shale combustion than from the conversion of peat into energy.

Health effects of indoor combustion of biomass The use of fuels such as wood, straw or dung to provide household energy for light, cooking and heating can create problems of indoor pollution. As illustrated in Table 7.5, without proper ventilation and especially if the stove is lacking a flue, relatively high levels of CO, CO_2, water vapour, SO_x, NO_x and particulate matter can accumulate in the home. In addition, PAHs, products of incomplete combustion, may be formed (Chadwick, 1990).

The major health effects stemming from the household use of biomass as a fuel have not been widely studied and are, therefore, not well understood, but are thought to be similar to the effects of both active and passive tobacco smoke (Smith, 1990). In addition to the risk of burns, potential health problems include acute respiratory infections (children are particularly susceptible), chronic obstructive lung diseases, low birth weights, cancer (inconclusive) and eye ailments (ancedotal evidence but no systematic study) (Smith, 1987).

Improvements in stove design have generally concentrated on fuel efficiency, although smoke reduction may be an additional benefit when combustion efficiency is maintained or improved along with thermal efficiency. An increase in thermal efficiency alone may actually increase pollutant emissions (Smith, 1990).

Table 7.4 *Health risks of low-grade fuel combustion for power generation (data in fatalities/GW(e))*

Fuel	Nature of risk	Occupational	Public
Coal	immediate	0.09–0.2[a], 0.03–0.44[b], 0.023[c]	
	delayed		0.77[a,d], 0.26[b,e,f], 3.2–22[g,h]
Peat	immediate	0.09[i], 0.05[j]	0.9–1.9[j,k]
Oil shale	immediate	0.013–0.02[b,l]	
	delayed		0.13[b,m], 0.0004[i,n,o], 0.57[i,n,p]

[a] Hamilton, 1984

[b] UNEP, 1985b

[c] Hubert et al., 1981

[d] Estimates from mortality from sulphate in the form of acid aerosols

[e] Morris, 1979

[f] Premature deaths: the estimate is based on the assumption that for 0.5 kg SO_2 emissions per 1.05 GJ input, the death estimate is 1.7 per plant per year per million within an 80 per cent confidence interval of 0–10; the base power plant is assumed to have a 305-m-high stack

[g] Black and Niehaus, 1980

[h] Stack height = 305 m; 2.2 million people within a radius of 80 km

[i] Hamilton, 1990

[j] Lautkaski et al., 1982

[k] Dust-burning conversion technology, 90 per cent peat and 10 per cent oil-burning; the information on the conversion stage is based on a 150-MW(e) power plant

[l] Assuming an oil-fired plant

[m] House, 1983

[n] IWG, 1982

[o] Excess cancer deaths due to airborne carcinogens; not additive to excess deaths due to exposure to sulphates

[p] Excess deaths from exposure to sulphates; not additive to excess deaths due to airborne carcinogens

(Source: Compiled from IAEA, 1991)

Table 7.5 Indoor air pollution from biomass combustion in developing countries

Location	Households	Duration	PM (mg/m³)	BaP (ng/m³)	CO (ppm)	Other
Nigeria, Lagos	98	?	–	–	940	NO_2: 8.6 ppm SO_2: 38 ppm Benzene: 86 ppm
Papua, New Guinea,						
–western highlands	6	All night	0.36	–	11	HCHO: 0.67 ppm
–eastern highlands	3	All night	0.84	–	31	HCHO: 1.2 ppm
Kenya, highlands	5	?	4.0	145	–	BaA: 224 ng/m³ Phenols: 1.0 µg/m³ Acetic acid: 4.6 µg/m³
Kenya, sea level	3	?	0.8	12	–	BaA: 20 ng/m³
Guatemala, two villages	180	?				
–poorly ventilated			–	–	26–50	
–well ventilated			–	–	15–31	
India, Ahmedabad						
–wood	5	15 min	7.2	1270	–	NO_2: 318 µg/m³ SO_2: 169 µg/m³

Table 7.5 (continued)

Location	Households	Duration	PM (mg/m³)	BaP (ng/m³)	CO (ppm)	Other
-cattle dung	4	15 min	16.0	8250	—	NO_2: 144 µg/m³ SO_2: 242 µg/m³
-dung plus wood	7	15 min	21.2	9320	—	NO_2: 326 µg/m³ SO_2: 269 µg/m³
India, Gujarat						
-Boria, a.m.	10	45 min	4.8	3550	—	—
-Boria, p.m.			8.2	3550	—	—
-Denapura, a.m.	11	45 min	2.7	2220	—	—
-Denapura, p.m.			4.3	3210	—	—
-Meghva, a.m.	10	45 min	4.9	6070	—	—
-Meghva, p.m.			10.0	2620	—	—
-Rampura, a.m.	5	45 min	6.2	5410	—	—
-Rampura, p.m.			5.6	3040	—	—
Seasonal conditions:						
-Monsoon conditions	1		56.6	19300	—	—
-Two-mouth chulha	1		14.0	4270	—	—
Nepal	22	8–12 h	606.0[a]	—		

[a] Respirable particulate matter

(Source: WHO, 1984; Smith, Aggarwal and Dave, 1983)

Particulates

Particulate matter is formed during combustion of low-grade fuels and includes dust and fly ash. It can range in size from fairly large visible particles to submicron particles. Efficient fuel combustion can decrease but not eliminate the quantities of particulate matter produced.

Environmental aspects of particulates

The greatest concern regarding particulates is their potential negative impact on human health. Very small particles, which are difficult to collect, may be inhaled and contribute to respiratory disease. Although most of the heavy metals in a fuel will remain as bottom ash, a portion may also be emitted into the atmosphere in particulate form leading to contamination of soil or surface water.

Control of particulate matter

Inorganic sulphur along with other fuel particulates including dust and fly ash can be captured through physical means by cyclonic devices, ESPs and fabric filters (baghouses). ESPs and baghouses are more effective than cyclones, although all methods remove at least 90 per cent of the particulate matter in flue gas. The choice of a particulate removal technology will depend, in part, on the fuel characteristics. For example, where fly ash has a high resistivity (e.g. low-sulphur coals), baghouses may be more effective than ESPs unless the ESPs are very large (EPRI, 1991).

Cyclones and multicyclones Cyclones are operated by using centrifugal forces in conical vessels to separate particulates from flue gas. Up until the 1950s they were the only form of particulate control (Sondreal, 1992) and are still commonly used to remove large particulates ahead of other pollution control devices. Multicyclones are also used alone in small combustors where emission standards are lenient (Smith et al., 1992). Cyclones are efficient at removing up to 85–95 per cent of particulate matter 10 μm and above although efficiency rates for particles 3–4 μm in size falls to less than 50 per cent (Seward, Hollis and Opalanko, 1978).

Electrostatic precipitators ESPs are a highly effective method of particulate removal (up to 99.5 per cent). In ESPs, particulate-laden flue gas is ionized when passed between plates 15–25 cm apart. Midway between each plate are wire electrodes. The charged dust particles then attach to the electrodes, emitting an opposite charge. This causes the particulates to become neutral and precipitate into the ash hopper. Some particulates will adhere to the surface of the electrode collectors and will need to be periodically removed to maintain the ionizing effect. ESPs are not effective for removing particles 0.5 μm or less in size.

Experiments in low-cost improvements of ESPs for high-resistivity ash and small (submicron) particles include intermittent energization of the electrodes, wide plate spacing, flue gas conditioning and two-stage precharging (Sondreal, 1992; Torrens, 1992).

Fabric filtration (baghouses) Fabric filtration involves the passage of flue gas through dense cloth filters which collect dust and fly ash. These cloth filters are traditionally known as baghouses and are made of synthetic material such as nylon or dacron or woven fibreglass. The filters must be cleaned from time to time. For this reason the bags are installed in a compartment formation so that dust removal and maintenance can take place with minimal disruption to the operating system. Baghouses have a high efficiency, generally removing 99.5–99.9 per cent of the particulates present in the flue gases. The drawbacks to their use include large space requirements. Like ESPs, baghouses remove few particles smaller than 5 μm.

The most recent improvement in fabric filtration is the pulse jet baghouse, which uses short pulses of high-pressure air to provide a force for cleaning that is more vigorous than conventional methods. In addition, the dust is collected on the outside of the filter, allowing cleaning of the baghouse while it is still on line (Torrens, 1992). A baghouse with reverse flow cleaning is illustrated in Figure 7.9.

Carbon dioxide

The organic component of all low-grade fuels is primarily carbon. During combustion the carbon combines with oxygen, which releases energy and produces CO_2. Through photosynthesis, when CO_2 is fixed, oxygen is released to the atmosphere, a process which is vital for the ongoing existence of plant and animal life on this planet.

Environmental aspects of CO_2

Over the last century increasing amounts of fuel, particularly carbon-dense fuels such as oil and coal, have been burned to provide energy for transportation, space heating and industrial development. This activity has resulted in large quantities of CO_2 being released into the atmosphere, where it is presently thought to contribute up to 50 per cent of the greenhouse effect (Patterson, 1990).

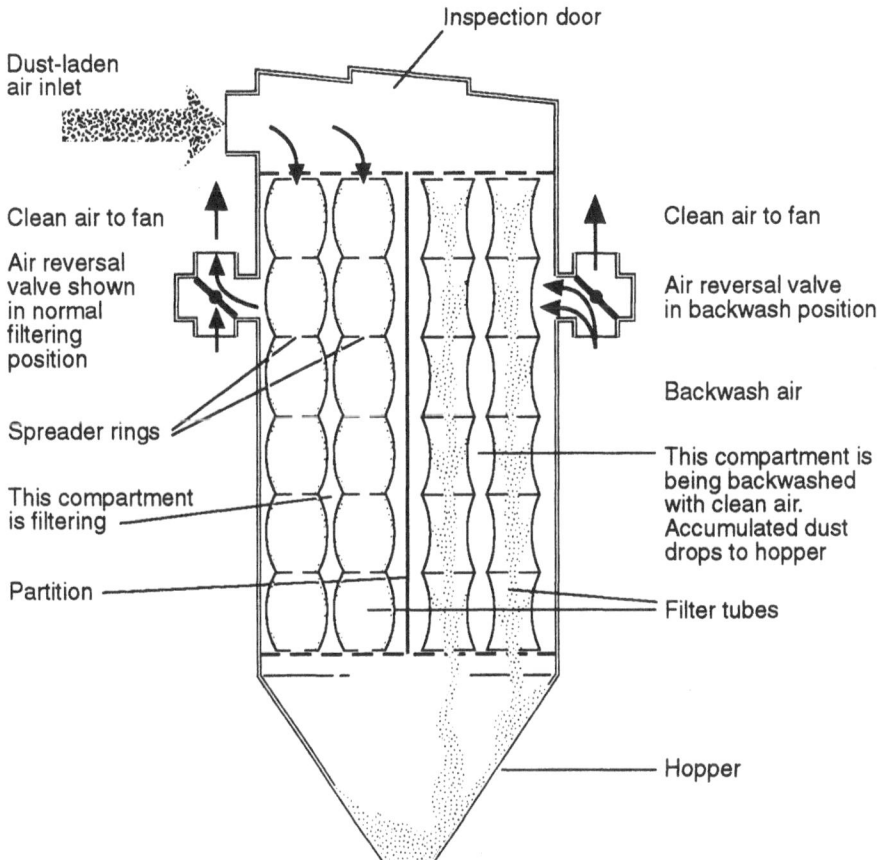

Figure 7.9 *Typical baghouse with reverse-flow cleaning*

(Source: US Department of Energy 1988; adapted from Argonne, 1978)

Environmental protection and CO₂

The technology does not exist to capture and contain the large quantities of CO_2 released during combustion; CO_2 emissions decrease only if less carbon is burned. This may be achieved through conservation measures and high end-use efficiency (see Chapter 8) and, in some cases, by switching to fuels such as natural gas, which produce less CO_2 per unit of energy produced.

Since 1991 Britain, France, Germany and Spain have been conducting research into methods for CO_2 removal as a 'fall back option in case increased efficiency proves an inadequate means of reducing CO_2 emissions' (IEACR, 1991). Investigations in this area are focusing on IGCC with a physical solvent system, chemical scrubbers at pulverized fuel plants and the operation of combustion systems using oxygen and recycled flue gas to produce a concentrated stream of CO_2 which can either be disposed of directly or subjected to scrubbing. The disposal of vast quantities of waste which would result from the collection of CO_2 is also an area currently being researched. FGD can increase CO_2 emission by decreasing efficiency (Patterson, 1992).

CO_2 production and low-grade fuels

Due to the composition of lignite and oil shale, greater quantities of CO_2 result from the burning of these fuels than from the use of peat, wood or biomass. Large-scale burning of biomass and wood is unacceptable, however, since these materials act as CO_2 sinks while they are growing and provide important functions in the ecosystems of which they are a part. The use of wood to meet industrial energy needs should only be attempted on a sustainable basis using wood from a plantation or wood wastes (e.g. from a pulp and paper mill). The CO_2 produced from the burning of MSW varies depending upon the contents of the MSW, particularly with respect to the quantity of plastics present in the waste. The emission of CO is lower from wood than from lignite combustion when an FBC system (including a cyclone) is used (Leckner and Karlsson, 1993).

Sulphur oxides

Fuel sulphur occurs in an inorganic (FeS_2) and organic form. Organic sulphur is part of the fuel macromolecules and is bonded with carbon in a coke structure or in compounds of mobile volatiles which separate from carbon during thermal decomposition. This volatile sulphur combines with oxygen to form SO_2 and a proportionately smaller quantity of sulphur trioxide (SO_3).

Environmental aspects of SO_x

Emissions of SO_x into the atmosphere pose a threat to the environment because sulphur mixes easily with water to form acidic compounds such as sulphuric acid and sulphates. Rain or snow laden with these compounds is known as acidic precipitation or acid rain. The corrosive properties of acid precipitation can lead to damage to man-made structures and can alter the chemical balance in water and soils where the alkali levels are not high enough to neutralize the sulphuric acids. This can have a negative effect on plant growth and entire ecosystems, which may be sensitive to acidic deposition.

Control mechanisms for SO_x

Sulphur removal requires an understanding of both organic and inorganic forms of sulphur. Inorganic sulphur (pyritic sulphur) is a free molecule which is not bonded with organic matter. Pyritic sulphur molecules are large enough that they may be removed relatively easily through physical cleaning methods. The removal of organic sulphur is much more difficult.

The sulphur in organic sulphur is bonded with dense heterocyclic structures. To destroy these dense structures it is necessary to use chemical desulphurization methods, making use of very strong and reactive extraction solvents. Approximately 70 per cent of organic sulphur is present in heterocyclic form.

Sulphur oxides may be removed from combustion gases through the use of FGD or through physical containment of sulphur by baghouses or ESPs. Processes for FGD involve the use of chemicals dissolved in water (wet FGD) or chemicals added to combustion gases in dry form. An overview of SO_2 removal efficiencies for various techniques is presented in Table 7.6.

Table 7.6 *Sulphur removal efficiencies of FGD technologies operating in the United States*

Process	Removal, %
Wet scrubbers	
Limestone without oxidation	60–96
Limestone with oxidation	84–94
Lime reagent	80–98
Alkaline ash	50–95
Sodium carbonate	80–95
Dual alkali	85–95
Dry FGD processes	
Lime spray dryers	60–95
Sodium spray dryer	70
Dry sorbent injection	70
Regenerable	
Wellman Lord	90–98
Magnesium oxide	92

(Source: Vernon and Soud, 1990)

SO_x production and low-grade fuels

Wood and, in most cases, peat have a very low sulphur content, so production of SO_2 is not an important environmental concern associated with their combustion. Lignite and oil shale vary greatly in composition from deposit to deposit, and the sulphur content of these fuels can range from less than 1 per cent up to 12 per cent. For this reason, stringent SO_2 emission control is required when lignite and oil shale are burned to produce energy.

Sulphur recovery using wet processes

Wet sulphur recovery processes are the most commonly used FGD techniques.

Lime/limestone wet scrubber processes The vast majority of wet scrubbing systems installed in the world use lime or limestone as the alkali to neutralize the SO_2. In some cases, particularly with low-sulphur coals, the alkalinity of the fly ash is used to assist the lime or limestone. The resulting calcium sulphite/sulphate mixture is then disposed of in ponds or in disused mines. The water-limestone mixture is very difficult to dewater and usually forms a trixotropic solid commonly known as sludge. In order to make the sludge more manageable, forced oxidation may be used to convert the sulphite to sulphate, which is more easily dewatered. If the forced oxidation is not carried out, then other materials such as fly ash, cement and soil are used to stabilize the sludge.

Although the principle of limestone scrubbers appears straightforward, in practice many operating problems have been caused by an inadequate knowledge of the process chemistry. Figure 7.10 illustrates the major components of a limestone and lime sludge process.

Double-alkali process The double-alkali process is designed to absorb the SO_2 in some soluble medium and then treat the adsorber effluent with lime or limestone in a reactor outside the scrubber loop. The usual absorbent is an alkali. The simplest system uses sodium sulphite to remove SO_2 from the flue gas. Bisulphite solution from the scrubber is then regenerated with lime or limestone to produce limestone solids. The solids are separated, and the clear sulphite solution is returned to the scrubber. The double-alkali process is illustrated in Figure 7.11.

Once-through systems A once-through FGD process is one in which the liquid adsorbent is not recirculated. The advantage of this process is that the danger of scaling is eliminated since a clear solution or weakly diluted slurry can be used. The chief drawback is that large quantities of liquid are

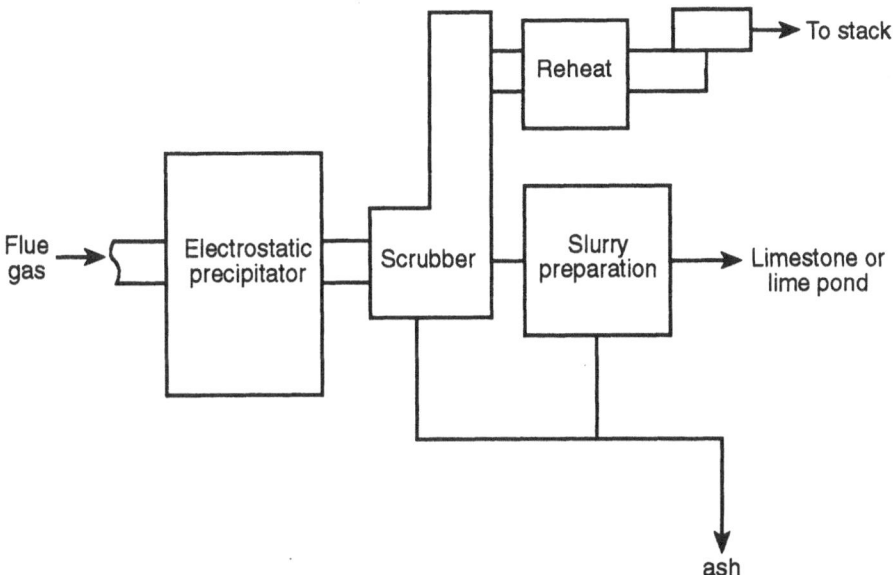

Figure 7.10 *Limestone and lime sludge process*

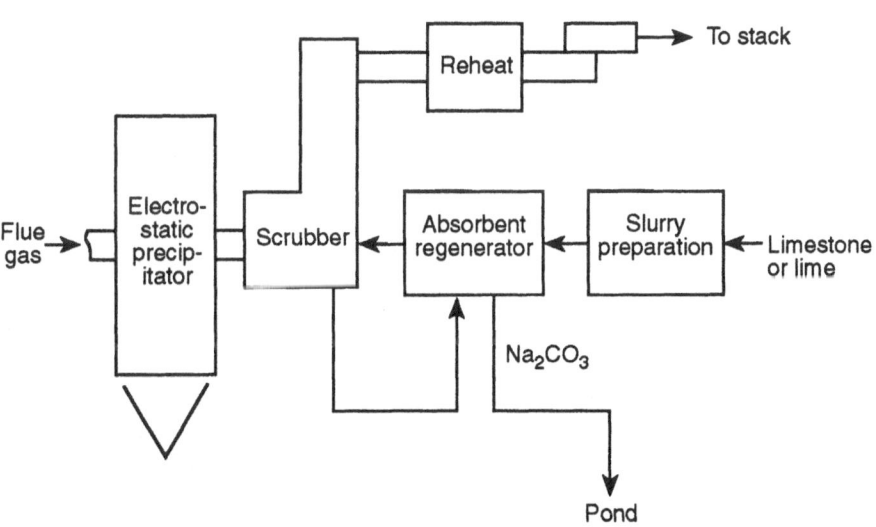

Figure 7.11 *Double-alkali process*

required, usually from a source such as a river or sea, and that flue gases are cooled to the temperature of the scrubbing water. This can lead to severe loss of buoyancy of the plume and under certain conditions can cause the plume to drop to ground level. The plume is also highly visible due to condensation droplets which are not evaporated.

Gypsum process Since the disposal of sulphite/sulphate sludge is problematic, investigation has taken place into the production of a more amenable product from FGD. Any of the wet scrubbing processes outlined above can be made to produce gypsum, a product used in the manufacture of building materials such as plasterboard.

Gypsum may be produced from FGD through the addition of an oxidation step in which calcium sulphite is oxidized to gypsum by air. Oxidation is usually carried out at a lower pH than that used for absorption. An illustration of the limestone and lime gypsum processes is given in Figure 7.12.

Dry alkali injection into the combustion chamber

The dry injection method involves the use of an alkali (usually slaked lime or limestone) which is injected into the hot boiler gases, where it reacts with the sulphur oxides and is then removed after reaction. Equipment for injecting the alkali into the flue gas and for removing it is required.

From 5 to 15 per cent of the SO_2 in the flue gases reacts with the alkaline elements already present in the lignite, oil shale or MSW, and the sulphur is then fixed and removed with the ash. The proportion of sulphur removed in this way depends upon the amount and alkalinity of the ash present in the fuel.

The chemistry of alkali injection Sulphur dioxide and SO_3 can be fixed through the injection of finely ground alkali and alkaline earth metals such as limestone ($CaCO_3$) and lime (CaO) directly into the combustion chamber. The added limestone must be ground to a very fine consistency, since coarse particles would cause a bridging of the limestone pores and hamper access of SO_2 to the remainder of the unreacted limestone surface.

The interaction between $CaCO_3$, or CaO, and SO_x will produce calcium sulphate ($CaSO_4$), also known as gypsum, and calcium sulphite ($CaSO_3$). In the case of oxidant shortage, the sulphur-fixing product is calcium sulphide (CaS), which is removed together with the slag by cyclone devices.

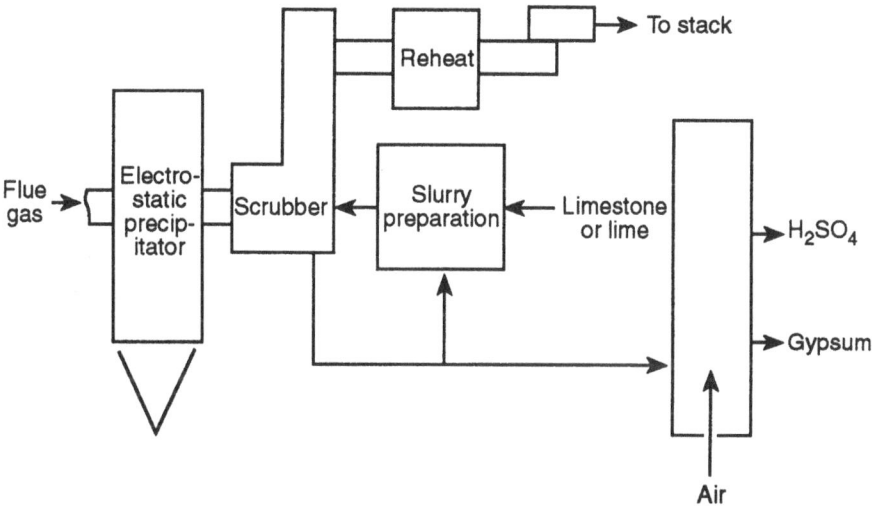

Figure 7.12 *Limestone and lime gypsum process*

Reactions The reactions which can occur when limestone is added to the flue gases are numerous. The following calcination, adsorption and oxidation reactions are significant:

$$CaCO_3 \longleftrightarrow CaO + CO_2$$
$$CaCO_3 + SO_2 \longleftrightarrow CaSO_3 + CO_2$$
$$CaCO_3 + SO_2 + \tfrac{1}{2}O_2 \longrightarrow CaSO_4 + CO_2$$
$$CaO + SO_2 \longleftrightarrow CaSO_3$$
$$CaO + SO_2 + \tfrac{1}{2}O_2 \longleftrightarrow CaSO_4$$
$$CaSO_3 + \tfrac{1}{2}O_2 \longleftrightarrow CaSO_4$$

Temperature considerations The temperature stabilities of the various species of calcium and magnesium are shown in Table 7.7. As Table 7.7 illustrates, the effective temperature range for reaction of the oxides is 660 to 1130 K for magnesium and 1045 to 1480 K for calcium. At temperatures lower than these the carbonates cannot react directly with the SO_2.

Table 7.7 *Maximum temperatures (K) for compound stability in flue gases (2000 ppm SO_2, 15 per cent CO_2, 1 per cent O_2)*

Absorber	$M(OH_2)$	MCO_3	MSO_3	MSO_4
Calcium	680	1045	729	1480
Magnesium	373	660	522	1130

The injection of the limestone with the fuel will result in a flame temperature of 2000 to 2300 K, which is higher than the decomposition temperature of calcium sulphate and also leads to 'dead-burning' of the lime. This is caused by the CaO forming a eutectic mixture (two or more substances, with the same minimum melting point, that behave in a similar way to a pure compound) with uncalcinated carbonate. This mixture then melts and forms a very non-porous structure which inhibits subsequent diffusion of SO_2 into the solid particles.

Applications of dry alkali injection High temperatures and a short residency time can lead to low efficiency of limestone and fouling problems in the boilers of fixed bed and pulverized fuel boilers. The removal efficiencies of limestone injection into conventional combustion systems have averaged between 20 and 40 per cent.

In FBC where the temperature is much lower (1000 to 1100 K) and the residence time of the fuel and limestone is longer, removal efficiencies of up to 90 per cent of SO_2 have been achieved. Lime injection is also successful in lignite-fired boilers, where the top flame temperature is 1420 K and the gas residence time in the boiler is longer than with high calorific value coals. Any unused lime also helps to cement the ash.

Dry limestone injection combined with wet scrubbing In this process about 80 per cent of the stoichiometric amount of lime is added to about 40 per cent of the total gas flow. The solid product is collected in a bag filter. The solids from the filter are then passed to a conventional wet-washing process. The use of lime and the long residence time on the bag filters ensures a high conversion in the dry process. The lime is thus used more effectively by combining the two processes. The overall efficiency of SO_2 removal, however, is not as high as in conventional wet washing.

Spray drying With this method the spray drier replaces the absorber, but upstream from the dust collector, SO_2 is adsorped in the droplets created by the drier's atomizer. The absorbent can be either lime, sodium carbonate or bicarbonate. A mixture of fully reacted and unreacted absorbent is produced which is then collected in an ESP or fabric filter together with the fly ash (Figure 7.13).

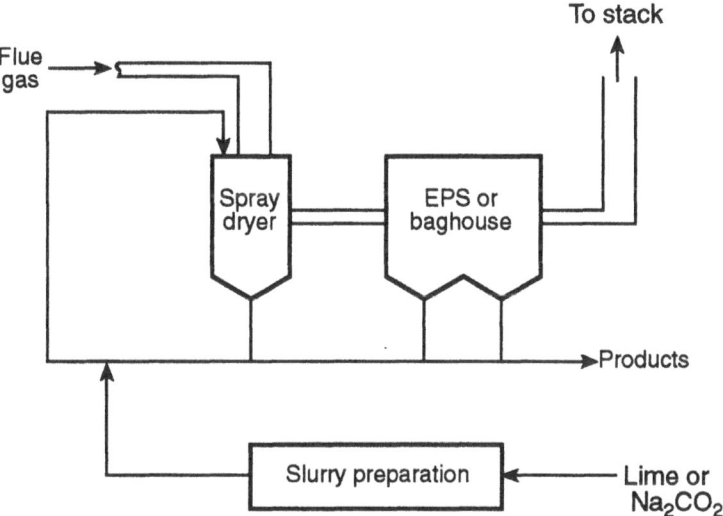

Figure 7.13 *The spray-dry process for sulphur removal*

The advantages of spray absorption include minimization of reheat, no scaling or plugging, the production of a totally dry product and low investment. However, large quantities of lime or other alkali must be imported to the site, and large quantities of product and unused reagent are produced which become thoroughly mixed with the fly ash and require careful disposal. The use of a reagent can be optimized by recirculating the ashes containing unreacted reagent. This method is less complex and less expensive than wet scrubbers for the high-calcium, low-sulphur coals.

Limestone injection multistage burners Interest in limestone injection into multi-stage burners has recently been revived due to investigation of this process for the reduction of NO_x emissions. Limestone injection multistage burners have lower flame temperatures, and because the introduction of air into the process is staggered, the flame is reduced. Sulphur-absorbing limestone is then injected into this fuel-rich reducing zone. The potential for limestone utilization is increased by the formation of calcium sulphide as a precursor to calcium sulphite.

Slagging combustor The slagging combustor is a device which can remove sulphur and particulate matter from low-grade fuels before it is injected into boilers or heaters. Although these combustors are primarily intended for retrofit applications, they will also be applicable and appropriate for incorporation into new facilities which can utilize their compact size and flexibility of fuel use. The fuel, either in a dry pulverized form or in a liquid mixture, is fed into the combustor in the conventional manner.

Dry adsorption

The cooling of flue gases caused by wet SO_2 scrubbing methods led to increased research into dry methods in the 1960s. The avoidance of reheat and the possibility that the SO_2 could be converted into a saleable product made these processes very attractive. The developments followed two main routes: chemisorption of SO_2 by metal oxides followed by thermal or chemical regeneration, and chemisorption on carbonaceous materials which were then regenerated by washing or heating.

Use of metal oxides

The alkalized alumina process is typical of metal oxide methods and has been studied extensively. One of the main difficulties in using solid adsorbents is that very large adsorbers have to be used to deal with the immense volumes of gas formed during combustion and to ensure adequate contact time for adsorption to occur.

The high cost of oxide adsorbent has made its chemical and physical integrity a primary economic factor. Cycling tests showed that the capacity fell to a steady level, but X-ray studies showed a loss of sodium probably caused by evaporation of sodium hydroxide formed as an intermediate in regeneration. The temperature and chemical cycling caused crystallite growth, which affected the diffusion characteristics of the adsorbent. It also weakened the pellets, causing uneconomically high losses from attrition.

Copper oxide The Shell copper oxide process overcomes some of the drawbacks of metal oxide use for sulphur adsorption by supporting the copper oxide on an alumina base and by using a parallel flow adsorber. In a fixed bed, since adsorption and regeneration are carried out at the same temperature, the adsorbent is not subjected to mechanical and thermal stresses.

The SO_2 removal efficiency reaches a maximum in about 20 minutes and then decreases as most of the copper oxide is converted to copper sulphate.

Carbon processes Activated carbon can adsorb SO_2 and it has been shown that, in the presence of water and oxygen, sulphuric acid will form on the surface of the activated carbon. In some instances, this phenomenon has been used to reactivate the carbon by water washing to produce weak sulphuric acid. The most common method for carbon reactivation is thermal regeneration at approximately 900 K. This thermal regeneration causes oxidation of the carbon surface with a consequent increase in the surface area and an increase in the sorptive capacity. The benefits of increased sorptive capacity are offset by the friable nature of the carbon at this stage and the increased risk of fire within the reactor.

Catalytic oxidation The most successful catalytic process is the Monsanto cat-ox process, which uses a vanadium catalyst. The catalyst is designed to operate at a temperature of approximately 720 K, where conversion of SO_2 to SO_3 is approximately 90 per cent. The SO_3 is then absorbed in sulphuric acid. The units are very large in this process because it is necessary to operate with a full gas flow.

Regenerable processes

These processes represent FGD at its most sophisticated. In these processes the absorbent is chemically or thermally reusable, and a saleable product (e.g. liquified SO_2, elemental sulphur or sulphuric acid) is produced. All regenerative processes produce by-products which result in a need for absorbent make-up and satisfactory by-product disposal.

Wellman-Lord process The Wellman-Lord process is based on the absorption of SO_2 in a concentrated solution of sodium sulphite and the regeneration of the resultant bisulphate (pyrosulphite) solution through steam stripping in an evaporator crystalliser. Figure 7.14 illustrates the basic layout of the Wellman-Lord process.

The crystallization of sodium sulphite in the evaporator regeneration is vital to the economic operation of the Wellman-Lord process. In the absorber, which is usually a valve-tray tower, SO_2 is absorbed into a concentrated sodium sulphite solution producing sodium bisulphite ($SO_2 + Na_2SO_3 + H_2O$ ---> $2NaHSO_3$).

Aqueous carbonate process The aqueous carbonate process, widely applied in the pulp and paper industry, uses an aqueous solution of sodium to adsorb SO_2 from the flue gases. The product of dry scrubbing is treated to regenerate the scrubbing and to produce elemental sulphur. A spray dryer

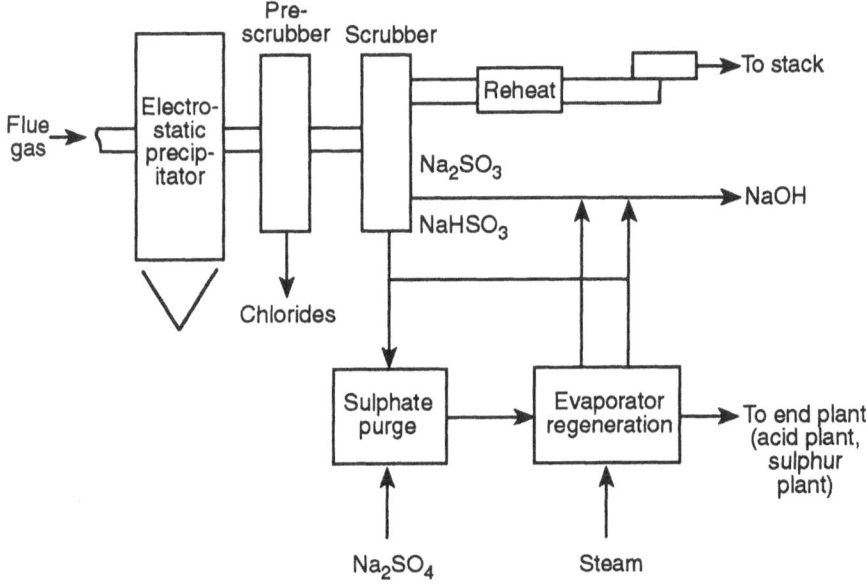

Figure 7.14 *Wellman-Lord process*

is used as an SO_2 absorber, which produces a dry granular salt mixture that is then reduced in a molten salt bath. The process can be described by the following reactions:

Absorption:

$$SO_2 + Na_2CO_3 \longrightarrow Na_2SO_3 + CO_2$$
$$Na_2SO_3 + \tfrac{1}{2}O_2 \longrightarrow Na_2SO_4$$

Reduction (at temperatures of 1170 to 1320 K):

$$2Na_2SO_3 + 3C \longrightarrow 2Na_2S + 3CO_2$$
$$Na_2SO_4 + 2C \longrightarrow Na_2S + 2CO_2$$
$$C + O_2 \longrightarrow CO_2$$

The CO_2 rich off-gas from the reducer is sent to the carbonization section. Reducer melt is continuously withdrawn and sent to the quench/dissolver vessel. The green liquor from the quench vessel is then exposed to the CO_2 rich reducer gas in order to regenerate the absorbent in the carbonization tower. The bicarbonate is heated to release CO_2 and the released hydrogen sulphide can be reacted in a Claus unit to form a saleable sulphur product.

Magnesium oxide process In the magnesium oxide process (Figure 7.15) an aqueous slurry of magnesium is used to absorb SO_2 ($MgO + SO_2 \longrightarrow$ $MgSO_3$). The resulting magnesium sulphite is then calcined to release SO_2 and to regenerate the absorbent ($MgSO_3 \longrightarrow MgO + SO_2$).

The SO_2 is available for conversion to either sulphuric acid or elemental sulphur. Since the magnesium salts are separated from the absorber as a solid and solid magnesium oxide is generated, it is possible to separate the absorption and regeneration units.

Nitrogen oxides

Nitric oxide (NO) and nitrogen dioxide (NO_2) are emitted from organic fuels during the combustion process. N_2O (nitrous oxide) may also be formed, particularly in very staged unstoichiometric processes. These two gaseous oxides of nitrogen are generally referred to as NO_x. There are two sources for the nitrogen in NO_x generated by combustion including nitrogen found in the oxidizing air and the nitrogen content of the fuel.

Environmental aspects of NO$_x$

Emissions of nitrogen oxides contribute to general levels of smog and air pollution. They also have an effect upon the acidity of acid deposition, since airborne NO_x can combine with water to form compounds such as nitrous acid and nitrates. The importance of NO_x in acid deposition is not yet well understood, although it is certain that it does play a role. Nitrous oxide is a highly stable compound which is formed during fuel conversion. It is similar to freons and other chlorofluorocarbons (CFCs) in that it has a long life in the troposphere, from where it may eventually move into the stratosphere and contribute to ozone depletion.

Control mechanisms for NO$_x$

The nitrogen present in the ambient air combines with oxygen at high temperatures to form thermal NO_x. This process is primarily dependent on the available concentration of oxygen and on the temperature of combustion. Nitrogen oxide formation of this type takes place at high temperatures, so the choice of combustion technologies with lower combustion temperatures such as fluidized bed gasification reduces the amount of NO_x produced.

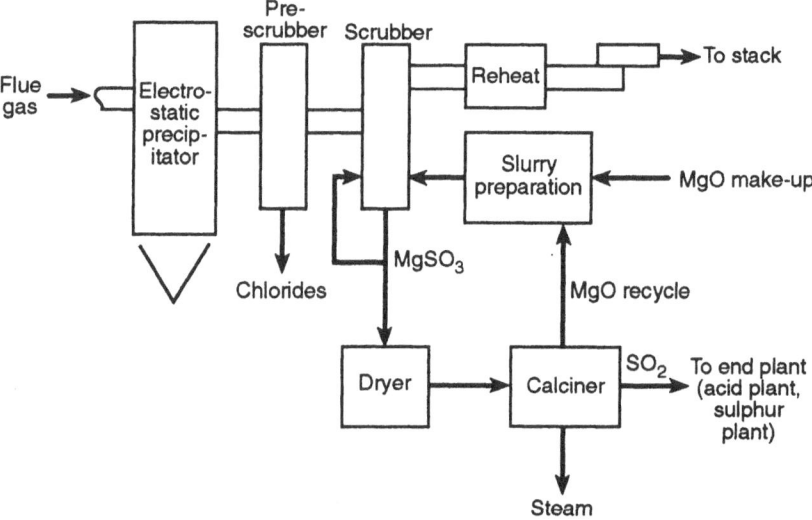

Figure 7.15 *Magnesium oxide process*

(Source: United Engineers, 1990)

However, N_2O emissions from all forms of FBC are higher than from conventional pulverized fuel systems (Sondreal, 1992).

The mechanism for the formation of NO_x from nitrogen present in fuel is not fully understood, so methods for suppressing NO_x in this way are still mainly experimental. There is some indication that this reaction is more likely at higher temperatures. Also, multistage combustion methods reduce NO_x formation, since the dimensions of the fuel particles decrease progressively through the stages, allowing a shorter release time for the fixed nitrogen. Mathematical modelling has been brought into play in the design of low NO_x burners using low-grade fuels (Fiveland and Latham, 1993).

Air staging and low NO_x burners are the most common approaches to controlling NO_x formation. If denitrification is required, a situation which may arise from a combination of fuel components and stringent legislation, then selective catalytic reduction (SCR) is the most frequently chosen method used with pulverized-coal utility boilers. An overview of the effectiveness for several NO_x control technologies is presented in Table 7.8.

Table 7.8 Effectiveness of NO_x control technologies

Control technology	Typical % reduction	Maximum reduction
Low-excess air	0–15	15
Biased firing	5	8
Burners-out-of-service	20–25	35
Overfire air	15–25	50
Low-NO_x burners	40–60	70
Thermal DeNO$_x$	30–60	70
SCR	50–80	90

(Source: US DOE, 1988)

NO_x and low-grade fuels

Since NO_x formation can occur from the nitrogen available in ambient air, the choice of conversion technology greatly influences the amount of NO_x produced by low-grade fuels. In addition, the nitrogen content of a fuel also contributes to NO_x formation. Generally, wood fuels and biomass have low nitrogen contents, as does black liquor from pulp and paper manufacture in Finland (Saviharju, 1990). The nitrogen content of peat, lignite and oil shale tends to fall within a higher range than other low-grade fuels. There is a negligible emission of NO_x if wood is combusted in a CFB system (Leckner and Karlsson, 1993).

Suppression of NO_x during combustion

Known methods of reduction and suppression of NO_x generated during combustion include reduction of combustion temperature, control of oxygen concentration, reduction of residence time in the high temperature zone, improvement of the devolatilization process and control of the air-fuel mixing ratio.

Load reduction It is well known that reduced load operation reduces the heat release per volume and per surface area and thus lowers peak flame temperature, resulting in lower thermal NO_x formation. Fuel-air mixing rates are also decreased, and this may lower NO_x emissions from fuel-bound nitrogen.

Minimum loads depend on burner and furnace designs and on fuel composition. For pulverized coal-fired units, turndown of 2–2.5:1 from full load is typical with all burners in service. Further load reduction is achieved by removing burners from service. For pulverized fuel boilers, NO_x emissions decrease almost linearly with load reduction.

Load reduction is an expensive proposition, since derating the boiler by 50 per cent essentially doubles the capital cost of providing power. The high cost of load reduction for a multiboiler system could be reduced by operating all units at slightly reduced load rather than operating some units at full load and removing others from service.

Low excess air combustion Boilers normally operate with excess air levels to promote efficient fuel combustion through complete burnout of fuel and avoidance of high carbon monoxide and smoke emission levels. As excess air decreases, the oxygen available in the flame zone decreases, and thus formation of both thermal and fuel NO_x is reduced.

Low excess air operation refers to operating the boiler at lower than normal excess air levels, usually close to smoke point. In addition to lowering NO_x emissions, low excess air operation reduces the gas flow rate and increases boiler efficiency. Consequently, NO_x control through low excess air operation can potentially result in cost savings.

Potential problems with this approach to NO_x suppression include increased smoke, CO and hydrocarbon emissions along with reduced combustion efficiency if excess air is not carefully controlled over the full boiler cycle. Increased furnace slagging may also occur at low excess air with certain types of coal.

Two-stage combustion Two-stage combustion involves operating some or all of the burners with less than the stoichiometric amount of air and supplying the remainder of the combustion air further downstream. This reduces the combustion temperature and suppresses NO_x formation. Additional air for full fuel oxidation is injected into the reheat zone by means of additional atomizers (Figure 7.16). The same effect is achieved during multistage combustion if oxidant feed is expanded to take place all along the combustion zone. This increases combustion time and encourages a more uniform temperature pattern in the flame. Furnace air staging has proven to be the most successful technology for meeting NO_x emission standards in Germany when using pulverized lignite as a fuel (Stratis, 1995).

Problems associated with two-stage combustion include corrosion and slagging. The water walls of conventional boilers are generally made of steel alloys which corrode under certain conditions. Under reducing conditions, the oxide layer which normally covers the tube surface is

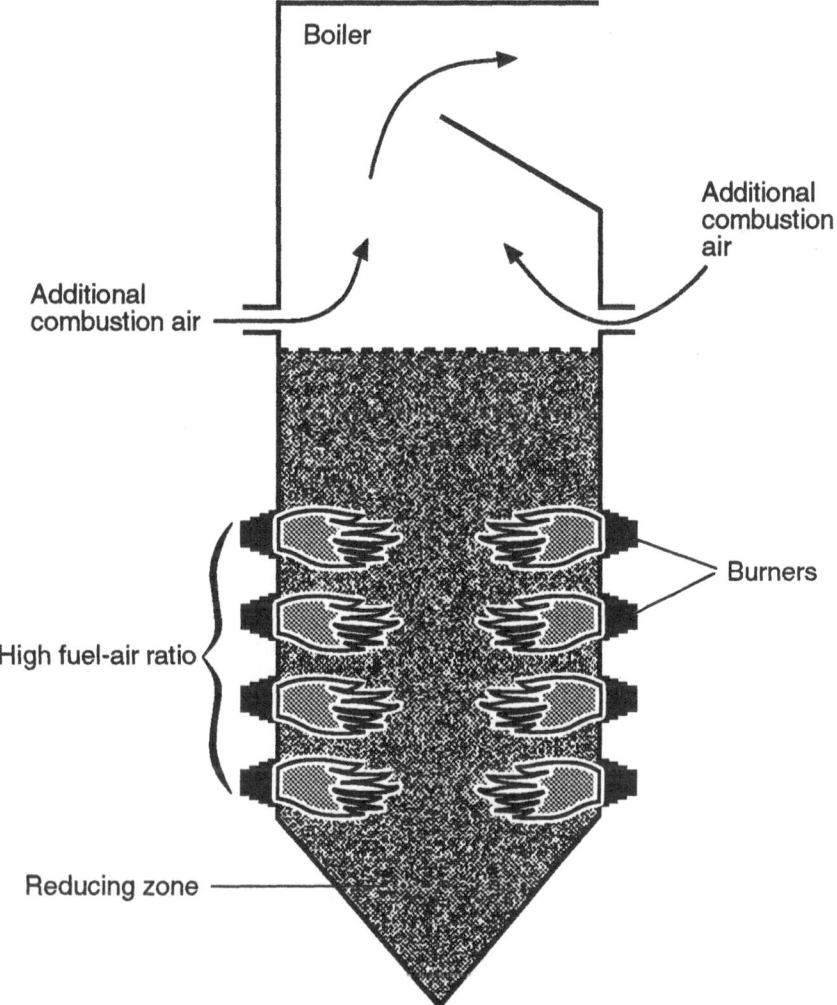

Figure 7.16 *Two-stage combustion for NO$_x$ reduction*

eliminated, resulting in accelerated corrosion. In addition, reducing conditions decrease the ash melting point, which can lead to slagging.

Flue gas recirculation Elimination of high-temperature zones in the furnace and levelling of the temperature pattern during combustion aid in the suppression of NO$_x$ formation. These conditions may be attained through flue gas recirculation and water or steam injection into the combustion zone. Figure 7.17 shows an arrangement of flue gas recirculation equipment on a typical wall-fired boiler. Since the burners are fired with partially diluted air, the concentration of oxygen at the base of the flame is reduced, and the overall flame temperature decreases. This has a pronounced effect on thermal NO_x emissions but only a slight impact on fuel NO_x. Therefore, flue gas recirculation is most effective with low-nitrogen fuels.

The cost of using flue gas recirculation to control NO_x emissions on retrofitted boilers can be considerable due to the requirements for new fans, dampers and dust collectors and the rearranging of boiler components. Flue gas recirculation is most effective during natural gas combustion. If low-grade fuels are combusted in a process using flue gas recirculation, then costs increase, since the use of natural gas must also be added.

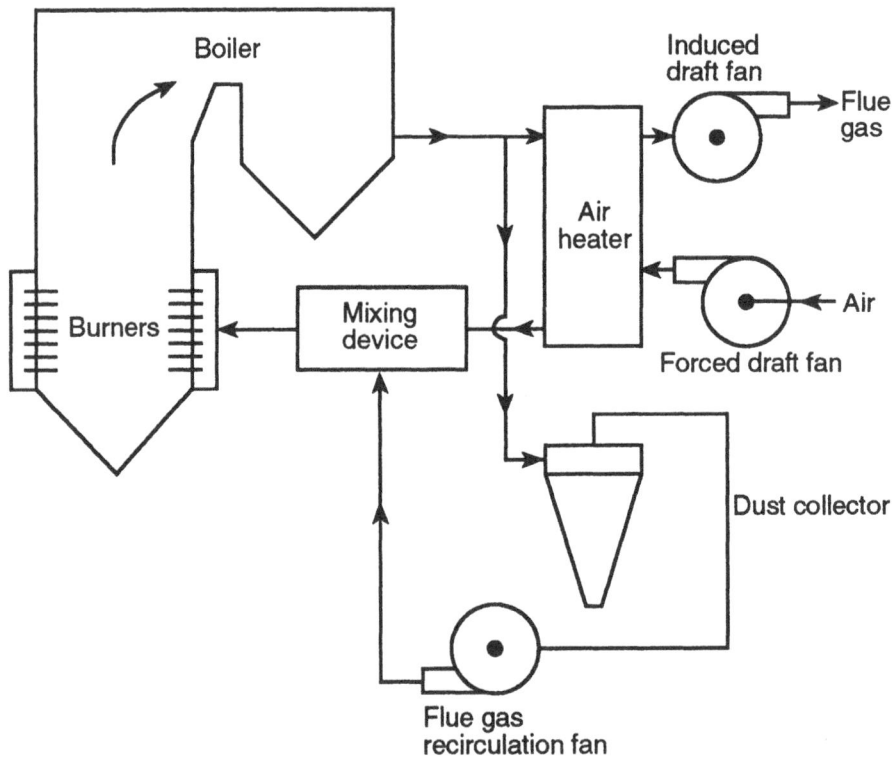

Figure 7.17 *Flue gas recirculation*

Reburning NO$_x$ may also be reduced by hydrocarbon species which may be present in a reacting flame. This concept is the basis for a process known as in-furnace NO$_x$ reduction or reburning.

Figure 7.18 illustrates this process as applied to a large utility boiler. Fuel is supplied to the furnace through two types of burners. The lower burners are fed approximately 90 per cent of the fuel and may be conventional or low NO$_x$ designs. The upper row of burners inject fuel but not combustion air. The remainder of the combustion air is injected above the burners. This arrangement allows the NO$_x$ produced by the low burners to react with the hydrocarbons injected through the upper burners. Testing with a Mitsubishi advanced combustion technology which makes use of reburning has achieved a 73 per cent reduction in NO$_x$ emissions compared with conventional firing (Aslanian, 1991).

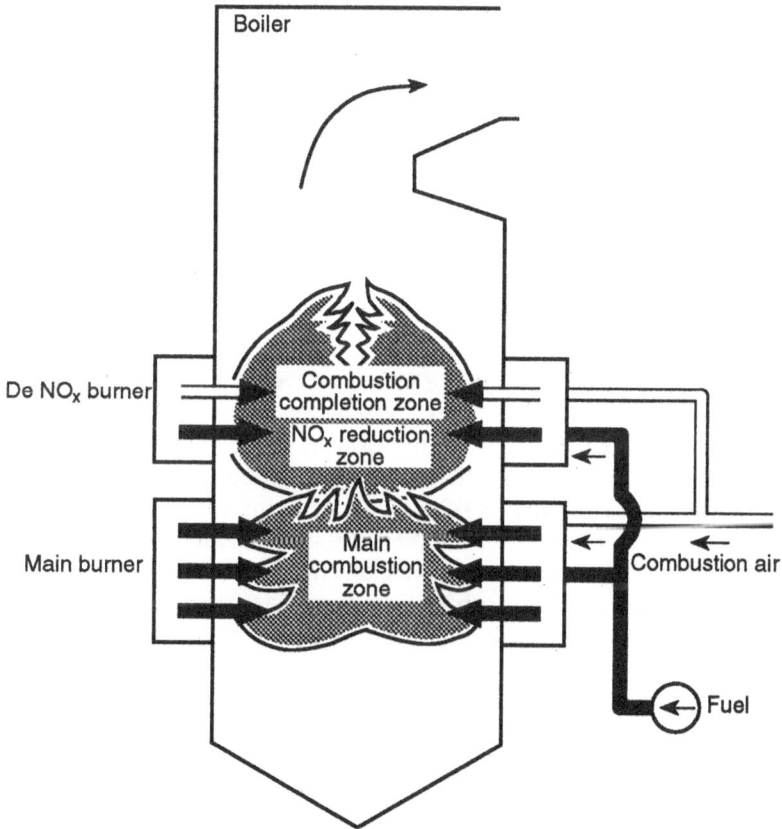

Figure 7.18 *Reburning: furnace NO$_x$ reduction*

Flue gas treatment for NO_x removal

If NO_x emissions cannot be controlled to the required levels through process or combustion modification, it is possible to remove NO_x using flue gas treatment systems. Nitrogen oxide removal of 60 to 90 per cent has been achieved through flue gas treatment, but it is significantly more expensive to construct, operate and maintain than combustion modification because it requires processing the entire volume of power plant flue gas.

Primarily due to high costs, flue gas treatment for NO_x removal has only been used commercially where NO_x emissions are very high (e.g. nitric acid plants) or where NO_x emission regulations require control beyond levels attainable through combustion modification techniques (e.g. Japan). Flue gas treatment systems developed especially for NO_x emissions from large thermal power plants or combustion sources with similar flue gas composition are dry processes which involve the reaction of NO_x with ammonia (NH_3) to produce nitrogen (N_2) with or without a catalyst.

Selective catalytic reduction In the SCR process NO_x emissions are reduced by reaction with NH_3 in the presence of a suitable catalyst such as titanium-vanadium. The reactions that occur are:

$$4NO + 4NO_2 + O_2 \longrightarrow 4N_2 + 6H_2O$$
$$6NO_2 + 8NH_3 \longrightarrow 7N_2 + 12H_2O$$

The optimal flue gas temperature range for these reactions is 300–400°C.

Pilot and full-scale evaluations of selective catalytic reduction processes have demonstrated that NO_x reductions in excess of 80 per cent can be routinely achieved with a high utilization of NH_3, although a number of problems exist with its application. Difficulties include a limited catalyst life, which means that nearly half the cost of using the process is spent on catalyst replacement and increased NH_3 emissions, resulting in the creation of odour problems. Greater emissions of NH_3 can also interfere with the effectiveness of a wet SO_x scrubbing system. The formation of $(NH_4)_2SO_4$ and NH_4HSO_4 may also result in fouling or plugging of downstream components.

Selective non-catalytic reduction In the selective non-catalytic reduction (SNR) process, NH_3 reacts with NO_x in the presence of O_2 at substantially higher temperatures than required in SCR. The NO_x reduction reactions are slow compared with typical combustion reactions. The use of NH_3 reactions to reduce NO_x emissions has been patented by Exxon (called Thermal $DeNO_x$) and has been applied to a wide variety of combustion-generated NO_x sources.

SNR is only approximately one-third as efficient as SCR in removing NO_x. For $NH_3/NO_3 = 1.0$, NO_x removal efficiencies are usually in the range of 25–37 per cent. SNR is similar to SCR with respect to problems with negative environmental effects of NH_3 emissions and component fouling caused by NH_4HSO_4. The cost of NH_3 is a primary determinant of the cost of using the selective non-catalytic reduction process for NO_x removal.

Halogens

The quantity and type of halogens emitted from a power plant are dependent upon the contents of a given fuel and on the approach taken to combustion and pollution control. The most common halogens found in low-grade fuels are chlorine (Cl), fluorine (F), bromine (B) and iodine (I).

Environmental aspects of halogens

Halogens react easily with many other substances and so, during combustion, can form corrosive compounds which may have a negative effect on the working components of a plant. In compound form halogens may contribute to the depletion of the ozone layer. Aromatic or polyaromatic polyhalogenated compounds such as polychlorinated dibenzo-*o*-dioxin (PCDD) and polychlorinated dibenzofuran (PCDF) may also be produced during combustion of chlorine, fluorine and bromine.

Control of halogens during combustion

Due to the high volatility of halogens, the greatest percentage end up in the flue gas following fuel combustion. In general, lower temperatures inhibit the release of volatile substances, so combustion technologies such as FBC can decrease the release of halogens compared with conventional fixed bed or pulverized fuel combustion methods.

Current flue gas-cleaning technology that makes use of reagents such as limestone or calcium hydroxide will reduce halogen emissions by greater than 95 per cent. When dry lime is used for FGD coupled with a baghouse filter, neutralization of the halogens will occur on the filter surfaces. Tests conducted on lignite with a high salt content by Finnish manufacturers of CFB boilers (Ahlström) showed that the addition of limestone and the use of baghouses captured close to 100 per cent of the HCl and HF produced during combustion (Hiltunen, 1990), although subsequent work has shown that the addition of iodine to CFBC can increase emissions of CO (Anthony, Bulewicz and Preto, 1993; Liang et al., 1991).

Halogens and low-grade fuels

Chlorine, fluorine, bromine and iodine have been found in all types of coal including lignite. Halogens are also known to be present in peat, oil shale, industrial wastes and sludges, and MSW (Frankenhaeuser and Hiltunen, 1994; Buckley and Domalski, 1988; Churney et al., 1988).

Polycyclic aromatic hydrocarbons

There are a number of compounds formed during fuel conversion which may prove to be a human health hazard or may affect other animals or plant life. One important group in this regard is polycyclic aromatic hydrocarbons (PAHs). The carbon compounds in this group number more than 600 and include dibenz(a,h)anthracene(DBahA),dimethylbenz(a)-anthracene, benzo(a)pyrene (BaP) and dibenzo(a,h)pyrene (DBahP).

PAHs are primarily formed and released during the thermal decomposition of fossil fuels. Under low-oxygen conditions decomposition of pyrolysis products can occur, resulting in low molecular weight volatile compounds which also promote the formation of PAHs. In general, there is a greater release of volatile organic compounds during FBC where temperatures are lower than during conventional combustion. Incomplete combustion resulting from the incompatibility of a boiler system and a fuel, a situation which may be more common for low-grade fuels, can also lead to increased formation of PAHs.

Environmental aspects of PAHs

The foremost concern regarding emissions of PAHs focuses on the carcinogenic or mutagenic effects associated with exposure to these compounds. Although BaP has been implicated as a likely mutagen and carcinogen, not all PAHs are carcinogenic (Chadwick, Highton and Lindman, 1987; Shabad, 1973). BaP may be present in emissions from power production and in the exhaust from all types of motorized vehicles. It may be inhaled or enter the food chain by settling on adjacent water bodies, soil and growing vegetables.

Control of PAHs

Large-scale and efficient power-generating plants produce relatively small amounts of PAHs, since it is generally possible to maintain a high enough combustion temperature so that these compounds will combine with oxygen and lose their pollutant potential. Other factors which reduce the formation of PAHs include the proper ratio of air and burning gases as well as an adequate residence time for the fuel and gas mixture in the active combustion zone. Poor maintenance or incorrect operation (e.g. allowing slag build-up on the combustion bed, which lowers the temperature of reduction) increases the quantities of PAHs produced. Experiments are ongoing with the University of North Dakota and Rheinbraun AG on the control of volatile organic carbons using activated char as a sorbent. Approaches include employing a char bed absorber or introducing powdered char into the system prior to the baghouse (Sondreal, 1992).

Low-grade fuels and PAHs

Volatile organic compounds generally arise out of incomplete combustion of all fossil fuels. The quantity produced is not related as strongly to fuel type as it is to the combustion technology employed (Chadwick, Highton and Lindman, 1987; Cope and Dacey, 1984). The combustion of oil shale, however, may result in the production of more PAHs per unit of energy produced than other fuels due to its high organic compound content as mined (Attassi, 1992; UNEP, 1985b; EPA, 1981, 1971). Recent studies indicate the presence of polycyclic aromatic compounds in the pyrolysis residues of oil shale lignite (Okutan, 1994).

Solid wastes

The combustion or pyrolysis of fuel produces ash, which essentially comprises the inorganic components of a fuel, that is, those components of a fuel which will not burn. The ash may fall out or be left behind following combustion (bottom ash/boiler slag) or may become airborne and subsequently be trapped by ESPs or fabric filters (fly ash). FGD also produces a residue which requires disposal.

Environmental aspects of solid waste

The primary concern regarding solid waste from fuel combustion centres on the trace heavy metals contained in the ash. Trace elements are widely represented in ash from all fuels. Although over 60 elements have been identified in fly ash from coal (Cope and Dacey, 1984), not all trace elements pose a threat to environmental quality. There is no worldwide agreement on the hazardous properties of all elements. The trace elements generally considered to be the most toxic to plant and animal life include silver (Ag), arsenic (As), beryllium (Be), boron (B), cadmium (Cd), chromium (Cr), copper (Cu), mercury (Hg), nickel (Ni), lead (Pb), selenium (Se) and zinc (Zn) (Ewers and Schlipköter, 1991; Vancil, Parrish and Palazzolo, 1991; Wilmoth et al., 1991; Smith, 1987; Cope and Dacey, 1984).

There are potential negative health effects associated with the inhalation, ingestion and absorption through the skin of heavy metals (Hamilton, 1992, 1979; Lautaski et al., 1982) and for these reasons some countries (e.g. the United States) have strict guidelines and regulations regarding the disposal of these substances. If ash containing toxic trace elements is disposed of in a landfill or ash pond, every effort should be made to remove the metals prior to disposal (e.g. through sedimentation within a closed system) or to ensure that the disposal area is secure against leakage (e.g. install a clay liner).

The mobilization of metals from solid waste materials placed in a landfill or ash pond can occur in a variety of ways (Förstner, Colombi and Kistler, 1991) including:

- precipitation;
- evaporation;
- gas moving within the fill;
- infiltration of groundwater;
- uncontrolled exfiltration of groundwater; and
- chemical reactions taking place over time in the fill.

The general build-up of solid waste can pose an aesthetic problem even if the waste has been treated to remove trace elements and precautions have been taken to achieve safe storage. These environmental considerations are summarized in Table 7.9.

Environmental protection measures for solid wastes

The quantity of solid waste produced by fuel combustion can be reduced through the use of conservation strategies and efficient operation of combustion technology (see Chapter 8). One method of waste reduction involves the use of low-waste technologies such as CFBC or PFBC. In these systems fuel particles stay suspended in the combustion chamber for a relatively long time, giving ample opportunity for all burnable material to undergo combustion. Larger particles of airborne trace elements can be captured by particulate retention measures such as ESPs and baghouses, while smaller trace element particles may be removed during wet scrubbing. The heavy metals removed most effectively by scrubbers include chromium, arsenic and nickel (Vancil, Parrish and Palazzolo, 1991). The bottom ash,

Table 7.9 *Solid waste and low-grade fuel conversion*

Potential impact	Environmental protection measure
Build-up of solid waste	
■land requirements for storage and disposal of bottom ash or collected fly ash	■opt for as efficient combustion as possible to decrease the quantity and improve the quality of ash produced
■leaching of pollutant and hazardous compounds from disposed or stored ash	■protect groundwater, adjacent surface water and soil by careful preparation of storage or disposal site using clay liner and leachate monitors and collectors; separate out pollutant and hazardous elements or compounds (e.g. heavy metals or phenols) and treat or destroy to reduce potential negative environmental impact
■unsightly mounds of slag and ash	■screen from view where possible, making use of trees and berms

trapped fly ash and the residue from FGD must be treated to remove heavy metals prior to disposal.

Heavy metals removed from combustion ash may be used in industry and, depending upon fuel characteristics and the sorbent added, products from FGD may be utilized (see Chapter 2). Removal of heavy metals may be aided by forming a slurry with the ash and precipitating out the undesired elements (e.g. vanadium and vanadium compounds may be precipitated by iron hydroxide, arsenic and arsenic compounds by iron sulphate and lime). Figure 7.19 outlines some of the processing and disposal options for FGD end-products.

Solid wastes from combustion may be disposed of in ash ponds in slurry form or in landfill as dry ash. It has been estimated that 70 per cent of worldwide production of bottom ash/boiler slag and fly ash ends up in landfill (Adriano et al., 1980). In selecting a suitable disposal site, it is necessary to consider the hydrogeological, geological, zoological, topographical and social aspects of the disposal area. This assessment is

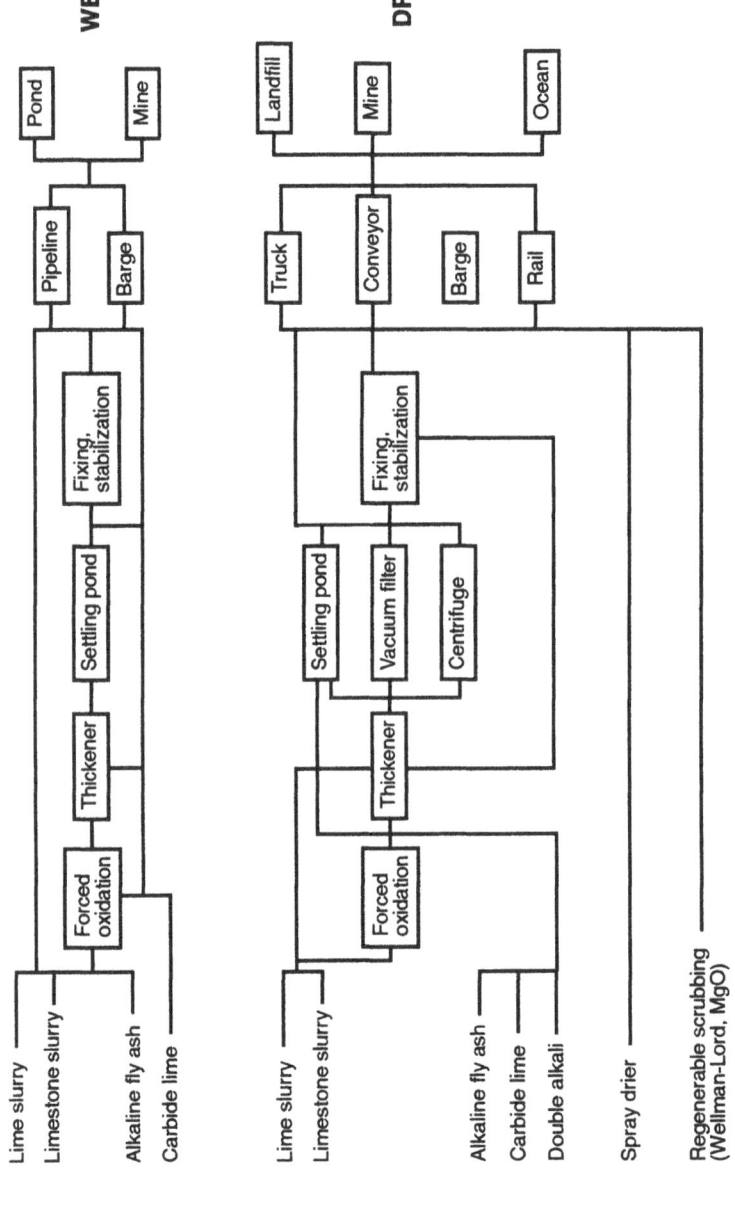

Figure 7.19 *Processing, transportation and disposal options for FGD and products*

(Source: Cope and Dacey, 1984)

required to prevent destruction of important ground cover and potential contamination of associated lands and ground and surface waters with heavy metals which may leach from the disposed ash. The best site would be in a natural clay basin, lined with additional clay, with installation of leachate monitors and pumps. If it is close to the combustion area and adequately prepared, ash may be returned to the original mine site for disposal.

Solid waste and low-grade fuels

Trace elements The heavy metal concentration in combustion products is dependent upon the metals content of the fuel. Metals are present in all fuels including low-grade fossil fuels, municipal waste and sewage sludge (see Chapter 1). As with all low-grade fuel characteristics, metals content can vary widely.

There is some evidence to indicate that the metals content of lignite is lower than higher-ranked coals, as mined (Swaine, 1990; Chadwick, Highton and Lindman, 1987; Heinrichs, 1983, 1977), although similar quantities of heavy metals may be present in the ash due to the lower energy density of lignite. Municipal waste and industrial sewage sludge generally have the highest metals content of all low-grade fuels (Häni, 1991; Vancil, Parrish and Palazzolo, 1991; Walsh, Pincince and Niessen, 1990).

Quantities of waste The quantities of waste produced by low-grade fuel conversion will reflect the mineral matter in the fuel, the processing and upgrading prior to combustion and the conversion technology. If fuel is gasified prior to combustion, the majority of solid waste will result from gasification rather than combustion. Lignite and oil shale combustion can produce large quantities of solid waste, while peat and wood may produce less due to greater moisture content.

Gaseous trace elements

Certain heavy metals are easily volatilized and therefore may also be present in flue gases (e.g. selenium, cadmium and lead) (Aittola, 1992; Meij, 1989). They may frequently be controlled by flue gas cleaning. Mercury, however, forms a special case, because it will vaporize rapidly even at relatively low temperatures and will escape into the atmosphere in gaseous form. The emission of mercury cannot be controlled by flue gas cleaning.

References

ADRIANO, D C, PAGE, A L, ELSEEWI, A A, CHANG, A C and STRAUGHAN, I (1980) 'Utilization and disposal of fly ash and other coal residues in terrestrial ecosystems: a review'. *Journal of Environmental Quality* 9: 333–344.

AHLSTRÖM (1990) *Pyroflow: CFB Technology.* Ahlstrom Pyropower, Helsinki.

AITTOLA, J P (1992) 'Emissions from low-grade fuel utilization'. In *The Environmentally Sound Management of Low-grade Fuels.* SEI, Stockholm.

ANTHONY, E J, BULEWICZ, E M and PRETO, F (1993) 'The effects of halogens on fluidized bed combustion systems'. In *Proceedings of the 1993 International Conference on Fluidized Bed Combustion.* American Society of Mechanical Engineers, New York.

ASLANIAN, G S (1991) 'Status report on clean coal technologies'. *Conference on Clean Coal Technologies.* UNESCO, Venice.

ASLHOCH, W, KELLER, I and HERBERT, P K (1990) 'The development of the HTW coal gasification process'. *Conference on Coal Gasification Power Plants.* EPRI, Palo Alto.

ATTASSI, M (1992) 'Environmental impacts of the exploitation of El-Lajjun oil shales in Syria'. In *The Environmentally Sound Management of Low-grade Fuels.* SEI, Stockholm.

BEER, J M and HOMOLA, V (1992) 'Innovative technological initiatives to upgrade power plants'. *Conference on Clean and Efficient Use of Coal: A New Era for Low-rank Coal.* IEA/OECD, Paris.

BLACK, S C and NIEHAUS, F (1980) 'Comparison of risks and benefits among different energy systems'. In *Interaction of Energy and Climate* (ed by W Bach, I Pankrath and J Williams). Reidel, Dordrecht.

BOYD, T J (ed) (1994) *Fluidized Bed Projects and Technology.* American Society of Mechanical Engineers, New York.

BRANWELL, A (1991) 'Perspectives on energy in eastern Europe'. *Conference on Problems of Energy Conservation Technologies and Environment Protection.* Varna, Bulgaria.

BUCKLEY, T J and DOMALSKI, E S (1988) 'Evaluation of data on higher heating values and elemental analysis for refuse derived fuels'. In *Proceedings of the 1988 National Waste Processing Conference.* American Society of Engineers, New York.

CHADWICK, M J (1990) 'Air pollution'. In *Bioenergy and the Environment* (ed by J Pasztor and L A Kristoferson). Westview Press, Boulder.

CHADWICK, M J, HIGHTON, N H and LINDMAN, N (1987) *Environmental Impacts of Coal Mining and Utilization.* Pergamon Press, Oxford.

CHIEWWATTAKEE, A C (1993) 'Reducing environmental impact from coal and lignite in Thailand'. In *IEA Second International Conference on the Clean and Efficient Use of Coal and Lignite.* Hong Kong.

CHURNEY, K L, LEDFORD, A E, BUCKLEY, T J and DOMALSKI, E S (1988) 'Chlorine mass balance in the combustion of refuse derived fuel'. In *Proceedings of the 1988 National Waste Processing Conference.* American Society of Engineers, New York.

COPE, D R and DACEY, P W (1984) *Solid Residues from Coal Use Disposal and Utilization*. IEA Coal Research, London.

COUCH, G R (1992) 'Low-rank coal in eastern Europe: opportunities and constraints'. *Conference on Clean and Efficient Use of Energy: The New Era for Low-rank Coal*. IEA/OECD, Paris.

COULTHARD, E J, KORENBERG, J and OSWALD, K D (1991) *Co-firing Coal and Hospital Waste in a Circulating Fluidized Bed Boiler*. ASME, New York (Reprint).

DARLING, S L (1994) 'Design and experience with large-size CFB boilers'. In *Fluidized Bed Projects and Technology*. American Society of Mechanical Engineers, New York.

EIPPER, H (1990) *Non-hazardous Waste Incineration and Heat Generation*. OSC/Thyssen Engineering GmbH, Essen (Reprint).

EIPPER, H (1986) 'Experience in combustion of lignite and peat of low calorific value'. *UN Conference on the Utilization of Low-grade Coal*. Thyssen Engineering GmbH, Essen (Reprint).

EPA (1981) *Environmental Assessment: Perspective on the Energy Oil Shale Industry*. Environmental Protection Agency, Cincinnati.

EPA (1971) *Water Pollution Potential of Spent Oil Shale Residues*. Environmental Protection Agency, Cincinnati.

EPRI (1991) *Clean Air Technology R & D*. Electric Power Research Institute, Palo Alto.

EPRI (1983) *Evaluation of Peat as a Utility Boiler Fuel*. Electric Power Research Institute, Palo Alto.

EWERS, U and SCHLIPKÖTER, H -W (1991) 'Intake, distribution and excretion of metals and metal compounds in humans and animals'. In *Metals and Their Compounds in the Environment* (ed by E Merian). VCH, Weinheim.

FIVELAND, W A and LATHAM, C E (1993) 'The use of numerical modelling in the design of low NO_x burners for utility boilers'. *Combustion, Science and Technology* **93**: 53–72.

FÖRSTNER, U, COLOMBI, C and KISTLER, R (1991) 'Dumping of wastes'. In *Metals and Their Compounds in the Environment* (ed by E Merian). VCH, Weinheim.

FRANKENHAEUSER, M and HILTUNEN, M (1994) 'Emissions from co-combustion of used packaging with peat and coal'. *Chemosphere* **29**: 9–11.

GEISLER, O J, HEINDRICH, F and GONDERMAN, B (1990) *Operational Experience with a Fluidized Bed Fired Boiler Plant for the Combustion of Wastes*. Thyssen Engineering, Essen.

GENERATOR INDUSTRI (1985a) *Ljusdal Heating Station: Peat on a Sloping Chain Grate*. Generator Industri AB, Partille.

GENERATOR INDUSTRI (1985b) *Hudiksvall Heating Station: Peat and Wood Chips on a Sloping Chain Grate*. Generator Industri AB, Partille.

GÖTAVERKEN ENERGY (1991) *Götaverken CFB Gasifier*. Götaverken Energy AB, Stockholm.

HAJICEK, D, MANN, M, MOE, T and HENDERSON, A (1993) 'The effects of coal properties on CFBC performance'. In *Fluidized Bed Combustion.* Vol 1. American Society of Mechanical Engineers, New York.

HAMILTON, L D (1992) 'Health and environmental management of low-grade fuels'. In *The Environmentally Sound Management of Low-grade Fuels.* SEI, Stockholm.

HAMILTON, L D (1990) *Personal Communication.* Brookhaven National Laboratory, Upton.

HAMILTON, L D (1984) 'Health and environmental risks of energy systems'. *Risks and Benefits of Energy Systems.* IAEA, Vienna.

HAMILTON, L D (1979) Health effects of electricity generation'. *Symposium on Health Effects of Energy Production.* Chalk River Nuclear Laboratories, Ontario.

HÄNI, H (1991) 'Heavy metals in sewage sludge and town waste compost'. In *Metals and Their Compounds in the Environment* (ed by E Merian). VCH, Weinheim.

HEINRICHS, H (1983) 'Trace element discharge from a brown coal-fired power plant'. *Environmental Technology Letters* **3**: 127–136.

HEINRICHS, H (1977) 'Emissions of 22 elements from brown coal combustion'. *Naturwissenschaften* **64**: 479–481.

HERBERT, P K, LOEFFLER, J C and WECHSTER, A T (1988) *GFB Gasification of Waste Material.* Lurgi GmbH, Frankfurt (Reprint).

HILTUNEN, M (1990) 'SO$_2$, NO$_x$ and halogens emission control in pyroflow circulating fluidized bed boilers when using low-grade fuels'. In *VTT Symposium 108: Low-grade Fuels.* Technical Research Centre of Finland, Espoo.

HOUSE P W (1983) *Report EP-0093.* US Department of Energy, Washington.

HOUSE P W (1981) *Energy Technology and the Environment.* US Department of Energy, Washington.

HOWARD, J R (1989) *Fluidized Bed Technology.* Adam Hilger, Bristol.

HUBERT, P, MOATTI, J P, MACCIA, C and FAGNANI, F (1981) *Les Impacts Sanitaires et Ecologiques de la Production d'Electricité: le cas français.* Centre d'étude sur l'évaluation de la protection dans le domaine nucléaire, Fontenay-aux-Roses.

IAEA (1991) 'Comparative environmental and health effects of different energy systems for electricity generation'. In *Senior Expert Symposium on Electricity and the Environment: Key Issue Papers.* IAEA, Vienna.

IDRC (1986) *Energy Research: Directions and Issues for Developing Countries.* International Development Research Centre, Ottawa.

IEACR (1991) *FGD Installations on Coal-fired Plants.* IEA Coal Research, London.

ISHIKAWAJIMA-HARIMA LTD (1991) *Fluidized Bed Boiler.* Ischikawajima-Harima Heavy Industries, Tokyo.

IWG Corporation (1982) *Health and Environmental Effects Document on Oil Shale.* IWG, San Diego.

JANSSON, S A (1992) 'Status and development potential for PFBC plants'. *Conference on Clean and Efficient Use of Energy: A New Era for Low-rank Coal.* IEA/OECD, Paris.

JANSSON, S A (1990) 'Pressurized fluidized bed combustion of Spanish black lignite. Test experience, design and construction of a commercial scale PFBC in Spain'. *Low-grade Fuels.* Vol 1. Technical Research Centre of Finland, Espoo.

JI, C (1993) 'Coal and lignite: the needs and concerns of developing and industrializing nations'. In *IEA Second International Conference on the Clean and Efficient Use of Coal and Lignite.* Hong Kong.

JOHNS, P T, CLOCKER, R A and LEVSTEK, D F (1992) 'Modern utility boiler design for low-grade coals: an update'. *Conference on Clean and Efficient Use of Coal: The New Era for Low-rank Coal.* IEA/OECD, Paris.

JOHNS, P T, CLOCKER, R A and LEVSTEK, D F (1984) *Modern Utility Boiler Design for Low-grade Coals.* Babcock and Wilcox, Barberton (Reprint).

KAREKEZI, S, MARWICK, S, SIZOOMU, G and TURYAREEBA, P (1991) *Doing More with Less: Sustainable Development of the Wood Energy Sector in Uganda.* Kengo, Nairobi.

KIØRBOE, K G (1990) 'Danish experiences with combustion of low-grade fuels'. *Low-grade Fuels.* Vol 1. Technical Research Centre of Finland, Espoo.

LAUTKASKI, R, POHJOLA, V, SAVOLAINEN, I and VOURI, S (1982) 'A comparative assessment of the health impacts of coal-fired, peat-fired and nuclear power plants'. *Health Impacts of Different Sources of Energy.* IAEA, Vienna.

LEACH, G and MEARNS, R (1988) *Bioenergy Issues and Options for Africa.* IIED, London.

LECKNER, B and KARLSSON, M (1993) 'Emissions from circulating fluidized bed combustion of mixtures of wood and coal'. In *Proceedings of Twelfth International Conference on Fluidized Bed Combustion.* American Society of Mechanical Engineers, New York.

LIANG, D, ANTHONY, E, LOEWEN, B and YATES, D (1991) 'Halogen capture by limestone during fluidized bed combustion'. In *Proceedings of the Eleventh International Conference on Fluidized Bed Combustion.* Montreal.

LURGI (1988) *Oil Shale Retorting: The Lurgi-Ruhrgas (LR) Process.* Lurgi GmbH, Frankfurt.

MAKANSI, J (1991) Fluidized bed boilers. *Power* **135**: 15–32.

MANZOORI A R, LINDER, E R and AGARWAL, P K (1989) 'Transformation of inorganics during fluidized bed combustion of lignites'. In *Proceedings of the 1989 International Conference on Fluidized Bed Combustion.* San Francisco.

MEIJ, R (1989) 'Tracking trace elements at a coal-fired power plant equipped with wet flue-gas desulphurization facility'. *KEMA Scientific and Technical Reports, Special Issue* **7**: 267–355.

MORRIS, S (1979) 'Coal conversion technologies: some health and environmental effects'. *Science* **206**: 654–662.

MURPHY, M L (1994) 'Using fluidized bed boilers for burning refuse-derived solid fuel'. *Solid Waste Management Technology* **8**: 34–43.

OKUTAN, H (1994) 'Polycyclic aromatic compounds detected in pyrolysis residues of seyitomer oil shale and lignite'. *Environmental Technology* **15**: 333–342.

OTS, A A (1992) 'The use of Estonian oil shales for power generation'. In *The Environmentally Sound Management of Low-grade Fuels*. SEI, Stockholm.

OVEREND, R (1986) 'Bioenergy conversion process: a brief state of the art and discussion of environmental implications'. *Proceedings of International Union of Forestry Research Organization*. Ljubljana, Yugoslavia.

PATEL, J G and MENSINGER, M C (1990) 'Recent developments in gasification of low-grade fuels'. In *Low-grade Fuels*. Vol 1. Technical Research Centre of Finland, Espoo.

PATTERSON, W C (1992) 'Clean coal'. In *Emerging Energy Technologies*. Royal Institute of International Affairs, Dartmouth.

PATTERSON, W C (1990) *Coal Use Technology in a Changing Environment*. Financial Times Business Information, London.

PATTERSON, W C (1987) *Advanced Coal-use Technology*. Financial Times Business Information, London.

PETZEL, H-K (1994) 'Operating experience and perspectives in fluidized bed combustion in Europe'. In *Fluidized Bed Projects and Technology*. American Society of Mechanical Engineers, New York.

POERSCH, W and ZABESCHECK, G (1980) 'Fluidized combustion of fuels with different ash contents'. In *Fluidized Combustion: Systems and Applications*. Graham and Trotman, London.

REDDY, A K N (1985) 'An end-use methodology for development-oriented energy planning in developing countries with India as a case study'. *PU/CEES Report 181*. Princeton University, Princeton.

RIEDLE, K and BÖHM, B (1992) 'Overview of "best practice" technological options available for power generation, clean use of coal technologies and meeting environmental goals'. *Conference on Clean and Efficient Use of Coal: The New Era for Low-rank Coal*. IEA/OECD, Paris.

SAVIHARJU, K (1990) 'Combustion of low-grade fuels in Finland'. In *Low-grade Fuels*. Vol 1. Technical Research Centre of Finland, Espoo.

SEITZINGER, D L and MORRISON, J L (1993) 'Limestone selection for circulating fluidized bed boiler'. *Proceedings of the American Power Conference, Illinois Institute of Technology* **55**: 806–814.

SEWARD, W H, HOLLIS, J R and OPALANKO, R S (1978) *A Survey of Environmental Coal Use*. Argonne National Laboratory, Washington.

SHABAD, L M (1973) *Circulation of Carcinogens in the Environment*. Meditsina, Moscow.

SMITH, I M (1987) *Trace Elements from Coal Combustion: Emissions*. IEA Coal Research, London.

SMITH, I, HJARLMARSSON, A-K and SOUD, H S (1992) 'Environmental pollution control for power generation: an overview for low-rank coal'. *Conference on Clean and Efficient Use of Coal: The New Era for Low-rank Coal*. IEA/OECD, Paris.

SMITH, K R (1990) 'Health effects in developing countries'. In *Bioenergy and the Environment* (ed by J Pasztor and L A Kristoferson). Westview Press, Boulder.

SMITH, K R (1987) *Biofuels, Air Pollution and Health: A Global Review*. Plenum Press, New York.

SMITH, K R , AGGARWAL, A L and DAVE, R M (1983) 'Air pollution and rural biomass fuels in developing countries: a pilot village study in India and implications for research and policy'. *Atmospheric Environment* **17**: 2343–2362.

SONDREAL, E A (1992) 'Clean utilization of low-rank coals for low-cost power generation'. *Conference on Clean and Efficient Use of Coal: The New Era for Low-rank Coal*. IEA/OECD, Paris.

STRATIS, J A (1995) 'Chemometrical data treatment to study the environmental pollution around lignite power plants'. *Toxicology and Environmental Chemistry* **47**: 71–76.

SWAINE, D J (1990) *Trace Elements in Coal*. Butterworth, London.

THEIS, K A and LAMBERTZ, J (1990) 'State of development of the HTW process regarding its suitability for combined cycle power stations'. *Low-grade Fuels*. Vol 1. Technical Research Centre of Finland, Espoo.

TORRENS, I M (1992) 'Fossil power plant control technologies: status, performance, cost'. *Conference on Clean and Efficient Use of Coal: The New Era for Low-rank Coal*. IEA/OECD, Paris.

UHDE (1987) *Rheinbraun HTW Process*. Uhde GmbH, Bad Soden.

UNEP (1985a) *The Environmental Impacts of the Production and Use of Energy*. UNEP, Nairobi.

UNEP (1985b) *The Environmental Impacts of Exploitation of Oil Shales and Tar Sands*. UNEP, Nairobi.

UNITED ENGINEERS (1990) *Magnesium Oxide Flue Gas Desulphurization*. United Engineers and Contractors, Philadelphia.

US DEPARTMENT OF ENERGY (1988) *Energy Technologies and the Environment: Environmental Information Handbook*. US Department of Energy, Washington.

VANCIL, M A, PARRISH, C R and PALAZZOLO, M A (1991) *Emissions of Metals and Organics from Municipal Wastewater Sludge Incinerators*. EPA, Cincinnati.

VERNON, J L and SOUD, H N (1990) *FGD Installations on Coal-fired Plants*. IEA Coal Research, London.

WALSH, M J, PINCINCE, A B and NIESSEN, W R (1990) *Fuel-efficient Sewage Sludge Incineration*. EPA, Cincinnati.

WHO (1984) *Biomass Fuel Combustion and Health*. EFP 84.64. World Health Organization, Geneva.

WILMOTH, R C, HUBBARD, S J, BURCKLE, J O and MARTIN, J F (1991) 'Production and processing of metals: their disposal and future risks'. In *Metals and Their Compounds in the Environment* (ed by E Merian). VCH, Weinheim.

8

Low-grade fuel use efficiency and conservation

One of the primary reasons that industrially based economic development and environmental degradation are thought to go hand in hand is the great quantity of energy traditionally required to sustain economic growth. The profligate use of energy can also put a strain on a nation's financial resources, particularly if high-quality fossil fuels must be imported.

The benefits of using energy in an efficient manner are so numerous that it is difficult to believe that conservation and efficiency measures are not more widely implemented. Aside from a reduction in the need for energy imports, energy efficiency offers the potential for delaying the depletion of non-renewable natural resources. However, for a number of reasons, but particularly the relatively low price of oil on the international market, and the linkage between oil prices and the price of other fuels, conservation measures have only been partly effective. Energy conservation and efficiency strategies can play an important role in helping to meet energy demands in developing countries and regions where an adequate and reliable supply of energy is not always available to meet social and economic demands (Schweitzer, 1993).

From an environmental perspective the use of less energy is desirable because it would lower the quantity of pollutants entering the air, soil and water due to fuel combustion. Less energy production would potentially provide a positive impact throughout the entire fuel cycle. For example, residents in the vicinity of mining, upgrading facilities and combustion plants might experience less displacement or disruption if these activities were carried out on a smaller scale as a result of a decrease in energy demand.

This chapter provides guidelines for devising a conservation plan and outlines technical strategies for the energy-efficient use of low-grade fuels.

Energy conservation

Planning for energy conservation involves identifying goals and priorities for energy use as well as developing and implementing policies and programmes for reaching these goals. There are, however, problems with measuring the effectiveness of conservation measures. Energy is provided to meet an end-use demand. This demand can be reduced and potentially lead to a decrease in the amount of energy produced. Unfortunately, comprehensive data on end uses for energy are not available (Foley, 1991; Nørgård, 1991, 1989). This is due to the vast array of end uses and the traditional emphasis on supplying increasingly greater amounts of energy to support a growing economy.

Improved energy efficiency is often cited as a cornerstone of an effective conservation plan (Davison, 1992; Environment and Energy Institute, 1992; Bodlund et al., 1989; Daly and Cobb, 1989; Frisch, 1989; Flavin, 1986; Tata Energy Research Institute, 1983) . Certainly, producing more energy from less fuel could lead to a decline in energy use overall. The possibility exists, however, that the same amount of fuel (with the same quantity and quality of associated impacts) will be converted to energy in a more efficient way but will simply result in more secondary energy being used. In this instance, there would be no benefit in terms of fuel savings or environmental protection. For this reason a commitment to using less primary energy should be part of energy conservation planning. The major steps in developing and implementing a plan for enhanced energy efficiency are contained in Table 8.1.

Table 8.1 *Development and implementation of a conservation plan*

Required Action
1. Individual, private sector and public sector commitment to using less energy
2. Survey of existing energy use patterns and energy production equipment
3. Identification of specific conservation measures
4. Identification of constraints on implementation of a conservation plan along with measures for limiting or eliminating the constraints
5. Implementation of the plan
6. Monitoring and evaluation of conservation measures, revising plan as required

Developing a conservation plan

The first step towards energy conservation should involve the development of a conservation plan. To prepare a plan it is important to collect information that identifies energy needs on a sector-by-sector basis and options for meeting those needs in the most energy-efficient manner. This is commonly done through the use of an energy survey.

Conducting a survey An energy survey involves the use of detailed questions aimed at gathering information concerning the type and quantity of energy used, the technology used and the preferred pattern of energy use. Information may be difficult to collect due to the need to interview large numbers of people (this is especially true for surveys of domestic energy use) or foreign ownership of an industry that prevents access to the required figures. In carrying out a survey of this type, proper training of the interviewers (e.g. reviewing questions so that interviewers understand the meaning, explaining required format for answers received) is essential in order to ensure that the data are collected in a uniform and accurate manner.

Energy use patterns The breakdown of energy use among sectors can give a broad indication of the way in which energy is being used and therefore suggest an initial direction for energy efficiency measures. Regional power generation and industry will often be the greatest users of primary energy sources, often contrasted, especially in developing countries, with the agricultural and household sectors.

Energy use can be determined by social conditions such as population growth, settlement patterns (e.g. agricultural versus nomadic, urban versus rural) and traditional household cooking, lighting and heating practices. Information will also be required concerning the quantities and types of fuel available (the national energy balance).

Existing energy production equipment Over the last 25 years knowledge regarding efficient conversion of low-grade fuels has been increasing. Older equipment may often (though not always) be less efficient in energy conversion (Müller, Maher and Rath-Nigel, 1982). If the technology is operating in a reliable manner and money for capital expenditures is scarce, inefficient energy production may continue for the life of the energy technology (e.g. power plant or domestic stove). A survey of existing energy equipment will determine the capital outlay required for retrofitting, repowering, restoration or rehabilitation of power-generating units.

Identification of specific energy conservation measures There is a wide range of options for increasing energy efficiency which might be assessed with respect to an individual situation. Approaches to energy conservation centre on decreasing energy demand and using energy more efficiently. The latter can be achieved through improving the entire energy production process (e.g. improved maintenance of equipment), improving the efficiency of one component (repowering) or by investing in an entirely new process. These are discussed in more detail below.

Training in energy management and the technical aspects of energy efficiency (e.g. power plant maintenance) are essential for the implementation, monitoring and evaluation of conservation plans. Administration and the financial aspects of utility operation are also important.

Collecting knowledge of past experience in other regions concerning energy efficiency (e.g. effective policies and legislation, institutional frameworks) can be valuable in determining a workable conservation plan.

Constraints on the implementation of a conservation plan Impediments to energy conservation can come from a variety of directions including a lack of public and private sector commitment and political will, vested interests of the producers of primary energy, uncertainty regarding end-use demands, government policy, lack of energy-efficient products, capital scarcity, socially unacceptable ramifications and market issues (Schipper and Meyers, 1992). These must be mitigated or eliminated for a conservation plan to show positive results.

There appears to be a general belief that low prices for energy do nothing to encourage energy conservation. Eastern European countries in particular are being advised to make consumers pay the 'real price' of various fuels, partially to encourage energy efficiency, although many consumers in developed countries in Western Europe and North America still enjoy the use of subsidized energy.

Arguments for charging the true price of energy production have recently begun to include advocating the inclusion of environmental costs such as the cost of depleting a non-renewable energy source and the cost of environmental protection and clean-up in the price to consumers. Other issues in this realm include the role of tariffs, public monopolies and transnational corporations in preventing or encouraging energy conservation.

Energy-efficient power production and low-grade fuels

As emphasized above, a strong commitment to reducing the quantity of primary energy consumed and to increased end-use efficiency are essential for achieving greater levels of energy efficiency. From a technical perspective, energy efficiency measures for industrial and commercial applications of low-grade fuels generally involve three major approaches:

- overall energy system improvement;
- boiler improvement; or
- investment in an entirely new energy system.

Energy process improvement

The efficiency of an energy production system can be improved by ensuring that all components within the system are functioning to a high standard. This includes using fuel with specifications suited to the boiler installation and use of on-line diagnostic and performance monitors and good housekeeping practices.

Fuel suitability Using fuel with the wrong characteristics can severely impair the efficiency of an energy production plant due to problems such as slagging and fouling of the boiler fittings causing incomplete combustion and frequent system shutdowns. Processing of fuel prior to combustion can reduce the variability of the fuel entering the combustion chamber. This is particularly important when low-grade fuels are used in conventional pulverized fuel or grate boilers where fuel characteristics are highly related to boiler efficiency. FBC is often better suited to the conversion of low-grade fuels because the suspended nature of the fuel particles during combustion allows a wider variability in the characteristics of fuels used.

Diagnostic and performance monitors Diagnostic and performance monitors installed at various points throughout the system can increase efficiency by tracking any damage to components such as leaks in seals or corrosion due to wear. Performance monitoring generally refers to monitoring of the heat rate (energy produced per kilowatt hour) and can give an indication of any decline in thermal efficiency which might require attention.

The skill and training of personnel are very important in determining the success of a diagnostic and monitoring programme. Without adequate knowledge of the system workers will be unable to detect or do maintenance at the early stages of process breakdown.

Maximizing boiler combustion efficiency Boiler efficiency is dependent upon a number of factors, including size of the fuel particles, oxygen levels, pressure of combustion, size of the combustion chamber, presence of unburnt carbon, exit gas temperature, fan and pulverizer power, reheat spray flows, and main and auxiliary steam flows (Armor and Torrens, 1992; Huettenhain, 1992; Riedle and Böhm, 1992; Temchin and Feibus, 1992; Womack et al., 1992; Couch, 1990; IDRC, 1986; Rittenhouse, 1986; Johns, Clocker and Levstek, 1984).

A decrease in the net heat rate by changing the steam conditions in the boiler, if this is compatible with the existing system and will not cause an overload, is one example of improving efficiency of a pulverized fuel plant.

As Figure 8.1 illustrates, a 2 per cent savings in heat rate can be realized by increasing the pressure of the steam entering the turbine.

Identifying problems with the turbine Steam flow, pressure and temperature are important aspects of efficient turbine functioning. These can be affected by damage such as breaks or leaks in component seals or wear of component parts (e.g. nozzle, diaphragms, blades).

Good housekeeping practices In general, all problems should be dealt with as soon as they are recognized. Failure to repair leaks, wear or punctures can result in a loss of efficiency of production. For example, the escape of dust from a feeder is a loss of energy which could be avoided. Cleaning and sweeping throughout the plant on a daily basis also reduce the chance of accidents due to spontaneous combustion (Sondreal, 1992). This is especially important when using highly reactive low-grade fuels such as lignite.

Reduction in net heat rate

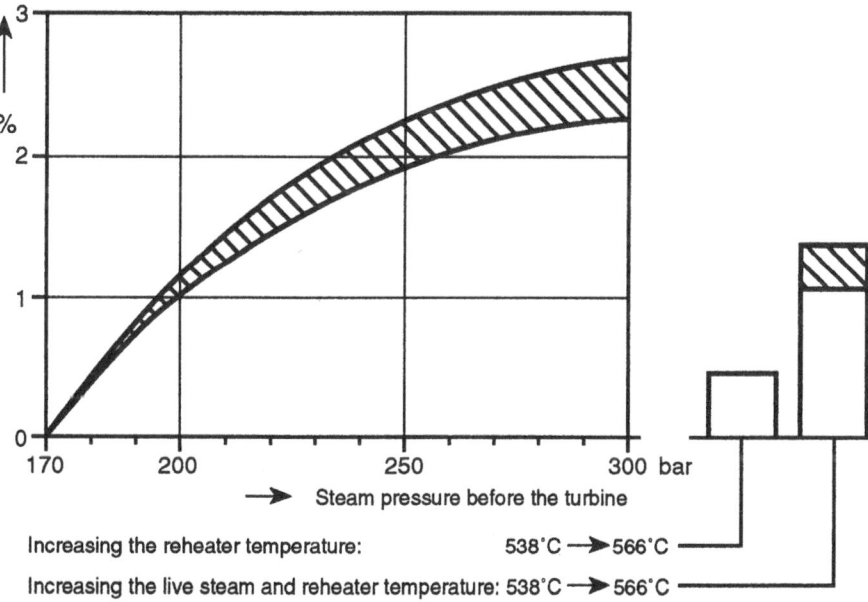

Figure 8.1 *Reduction in net heat rate by improving steam conditions*

(Source: Riedle and Böhm, 1992)

Boiler improvement

Boiler improvement through replacement (repowering) or repair is often carried out to extend the life of a power plant. Most approaches to boiler improvement also minimize the heat rate, thus increasing the efficiency of the entire generating unit, an example being boiler retrofit with gas turbines (Goss, 1992). In this case the exhaust from the gas turbine is used for steam generation in an unfired heat recovery steam generator. This can push heat efficiency to over 50 per cent (Beer and Homola, 1992; Emsperger, 1992).

Investment in a new energy system

The choice of a new combustion system must take into account the fuel type, the quantity of energy required, the space available for construction, environmental regulations and guidelines and efficiency goals. In general, different plants are optimized on the basis of different criteria. For example, increasing efficiency is usually a primary goal of energy production, but in the case of waste burning, where the control of pollutants may be more difficult, improving the quality of emissions may sometimes take precedence over increasing efficiency.

Pulverized fuel and grate boilers Traditional pulverized units that have come into operation over the last 20 years have a net efficiency of approximately 35 per cent including a scrubber. FGD technologies can decrease efficiency by as much as 4 to 5 per cent (Smith, Hjarlmarsson and Soud, 1992) but are required for environmental protection when using conventional boilers.

Fluidized bed combustion Atmospheric fluidized bed combustors have an efficiency comparable to conventional boilers, but they are generally better suited for use with low-grade fuels due to their ability to accept fuels with highly variable composition (Howard, 1989; Yaegar, 1984). Sulphur can be controlled through additives to the combustion chamber, and an AFBC operates at temperatures low enough to discourage NO_x formation.

Pressurized fluidized bed combustion Pressurized fluidized bed combustors have a high efficiency of 40 per cent (plus) which could be higher if inefficiencies in the cleaning of hot gas entering the turbine could be solved (Armor and Torrens, 1992; Beer and Homola, 1992; Jansson, 1992). These units are also suitable for use with low-grade fuels due to flexibility in accepting a variety of fuel characteristics and good environmental control. These are, however, expensive units and a relatively new technology. There are only a few plants in operation, although they have been proven to operate efficiently with lignite (Jansson, 1992).

Integrated advanced power systems These units can be integrated into an existing plant or may form the basis of a new plant. IGCC involves the use of a fully fired combined cycle where the gas turbine is combined with a fired steam generator. As Table 8.2 illustrates, these units have a high efficiency potential. The majority, however, are still in the demonstration stage.

Table 8.2 *Potential efficiencies of integrated advanced power systems*

Cycle	Potential efficiency based on higher heating value %
PFBC combined cycle	41–43
IGCC wet-gas cleaning	39–44
IGCC hot-gas cleaning	41–46
Potassium topping cycle (850°C)	48
Pressurized slag tap furnace (1350°C)	50–52

(Source: Martin, 1988)

Waste heat recovery

Energy use efficiency is more than doubled where the so-called waste heat from power production is utilized. The most common approach is to use the heat for space heating of plant buildings and, depending upon plant location, for regional space heating. This can be achieved by using heat exchangers, heat pumps or a thermal wheel which directs the heat from a gas tube containing cooler air.

Co-generation

Industries which produce steam for heating may use some of the steam to generate electricity for the plant. Known as co-generation, this can be both cost-effective and energy efficient. This process involves producing steam in a boiler, then directing the steam through a gas, steam or diesel turbine to manufacture electricity. In the United States low-grade coal has been used successfully to provide two separate products (steam and energy) to two different buyers (Perry and Lenertz, 1994). The use of a gas turbine also allows for space heating using waste heat (Shell, 1992). A co-generation process is illustrated in Figure 8.2.

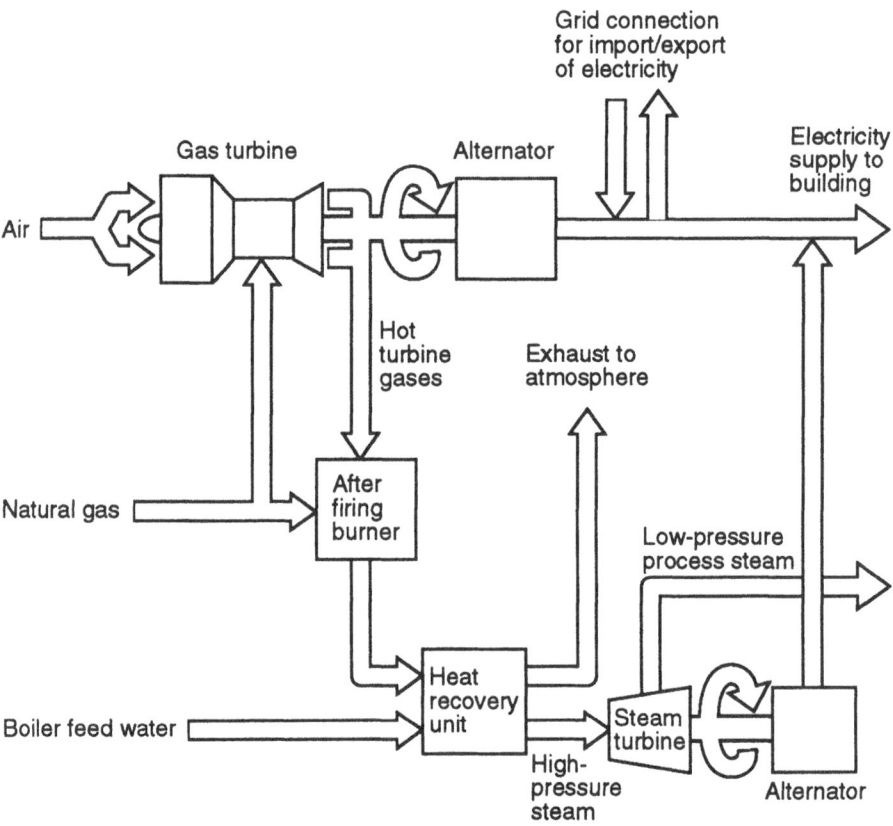

Figure 8.2 *Co-generation*

(Source: Shell, 1992)

References

ARMOR, A F and TORRENS, I (1992) *Coal-fired Power Plant Upgrades for Productivity and Environmental Performance*. EPRI, Palo Alto (Reprint).

BEER, J M and HOMOLA, V (1992) 'Innovative technological initiatives to upgrade power plants'. *Conference on Clean and Efficient Use of Coal: A New Era for Low-rank Coal*. IEA/OECD, Paris.

BODLUND, B, MILLS, E, KARLSON, T and JOHANSSON, T B (1989) 'The challenge of choices: technology options for the Swedish electricity sector'. In *Electricity*. Lund University Press, Lund.

COUCH, G R (1990) 'Lignite, low-grade coals and peat: an overview of resources power generation and upgrading'. In *Low-grade Fuels*. Vol 1. Technical Research Centre of Finland, Espoo.

DALY, H E and COBB, J B (1989) *For the Common Good*. Beacon Press, Boston.

DAVISON, A (1992) *Least Cost Planning: Should Utilities Invest in Energy Efficiency Rather Than in New Supplies?* Oxford Institute for Energy Studies, Oxford.

EMSPERGER, W (1992) 'Topping of existing and new power plants with gas turbines'. *Conference on Clean and Efficient Use of Coal: A New Era for Low-rank Coal*. IEA/OECD, Paris.

ENVIRONMENT AND ENERGY INSTITUTE (1992) 'Approval of UNCED resolution delayed by industry lobbying'. *Earth Summit Update* 7: 2.

FLAVIN, C (1986) *Electricity for a Developing World: New Directions*. Worldwatch Paper No. 70. Worldwatch Institute, Washington.

FOLEY, G (1991) *Energy Assistance Revisited: A Discussion Paper*. SEI, Sweden.

FRISCH, J-R (1989) *World Energy Horizons, 2000–2020*. IEA, Paris.

GOSS, W L (1992) 'Power plant retrofit of pollution controls'. *Conference on Clean and Efficient Use of Coal: A New Era for Low-rank Coal*. IEA/OECD, Paris.

HOWARD, J R (1989) *Fluidized Bed Technology*. Adam Hilger, Bristol.

HUETTENHAIN, H (1992) 'Low-rank coal upgrading technology review'. *Conference on Clean and Efficient Use of Coal: A New Era for Low-rank Coal*. IEA/OECD, Paris.

INTERNATIONAL DEVELOPMENT RESEARCH CENTRE (1986) *Energy Research: Directions and Issues for Developing Countries*. IDRC, Ottawa.

JANSSON, S A (1992) 'Status and development potential for PFBC plants'. *Conference on Clean and Efficient Use of Energy: A New Era for Low-rank Coal*. IEA/OECD, Paris.

JOHNS, P T, CLOCKER, R A and LEVSTEK, D F (1984) *Modern Utility Boiler Design for Low-grade Coals*. Babcock and Wilcox, Barberton (Reprint).

MARTIN, H (1988) 'Increase in process efficiency of coal-fired power plants'. *VGB Kraftwerkstechnik* 68: 199–205.

MÜLLER, M, MAHER, K and RATH-NAGEL, S (1982) *Energy Technology Systems Analysis Project: Summary Report on Technology Characterizations.* International Energy Agency, Paris.

NØRGÅRD, J S (1991) *Energy Conservation through Efficiency and Sufficiency.* Technical University of Denmark, Lyngby.

NØRGÅRD, J S (1989) 'Low electricity appliances: options for the future'. In *Electricity* (ed by T B Johansson, B Bodlund and R H Williams). Lund University Press, Lund.

PERRY, M A and LENERTZ, R P (1994) 'Cedar Bay cogeneration project: a plant in transition'. In *Fluidized Bed Projects and Technology.* American Society of Mechanical Engineers, New York.

RIEDLE, K and BÖHM, B (1992) 'Overview of "best practice" technological options available for power generation, clean use of coal technologies and meeting environmental goals'. *Conference on Clean and Efficient Use of Coal: The New Era for Low-rank Coal.* IEA/OECD, Paris.

RITTENHOUSE, R C (1986) 'Additives as a lower cost alternative to hardware retrofits'. *Power Engineering* **90**: 18–24.

SCHIPPER, L and MEYERS, S (1992) *Energy Efficiency and Human Activity.* Cambridge University Press.

SCHWEITZER, S C (1993) 'Environmentally sound technologies for sustainable development'. In *IEA Second International Conference on the Clean and Efficient Use of Coal and Lignite.* Hong Kong.

SHELL (1992) *Managing Energy Efficiently.* Group Public Affairs, Shell Briefing Service, Shell International Petroleum Company, London.

SMITH, K, HJARLMARSSON, A-L and SOUD, H S (1992) 'Environmental pollution control for power generation: an overview for low-rank coal'. *Conference on Clean and Efficient Use of Coal: The New Era for Low-rank Coal.* IEA/OECD, Paris.

SONDREAL, EA (1992) 'Clean utilization of low-rank coals for low-cost power generation'. *Conference on Clean and Efficient Use of Coal: The New Era for Low-rank Coal.* IEA/OECD, Paris.

TATA ENERGY RESEARCH INSTITUTE (1983) *Energy Conservation Bulletin.* Tata Energy Research Institute, Bombay.

TEMCHIN, J and FEIBUS, H (1992) *Methodology to Evaluate Alternative Approaches to Upgrade Power Stations in Eastern Europe.* US Department of Energy, Washington (Reprint).

WOMACK, E A, KITTO, J B, CLESSURAS, G J, KULIG, J S and LEVSTEK, D F (1992) 'Improving performance of power plants in central and eastern Europe'. *Conference on Clean and Efficient Use of Coal: The New Era for Low-rank Coal.* IEA/OECD, Paris.

YAEGER, K (1984) 'R & D status report: coal combustion system division'. *EPRI Journal* **9**: 43–48.

9

The transfer of environmentally sound technology for low-grade fuel use

Technology transfer is a broad subject which has been studied in depth from both a theoretical and a practical perspective. The issues briefly outlined here are those thought to be most closely related to a consideration of technology transfer and low-grade fuels.

In many instances the technology which is used for mining, processing and converting low-grade fuel is not the 'best available technology' either from the point of view of environmental protection (including fuel-use efficiency) or economic returns. A great deal of power production equipment is several decades old, and as long as it continues to meet demand it will not be upgraded or replaced until absolutely necessary due to the high cost of purchasing new equipment and the shut-down time required for procedures such as boiler retrofit. There is also the question of how to obtain newer technology when it is not an indigenous innovation or when funds are scarce.

Technological advancement, to date, has been one of the fundamental components of economic growth. This means that obtaining technology can be a high priority for less-developed countries in particular. The high costs associated with the research and development of energy-related technologies result in most of the work in this area being carried out by companies based in the richest nations. Thus, there is a growing chasm between countries which can obtain and make use of the latest technological developments and those which cannot.

Technology which is designed and built in one region may become available and used in another region through a variety of different avenues. Taken together these approaches are traditionally referred to as technology transfer, although other terms have been suggested in an attempt to clarify the meaning of the process.

The discussion below includes a consideration of various definitions of technology transfer, an overview of the controversy surrounding technology transfer, the primary approaches including mechanisms and constraints associated with obtaining technology and the issues associated specifically with the transfer of technology for use with low-grade fuels.

Towards a definition of technology transfer

A definition of technology transfer requires clarification of what is meant by both technology and transfer. There is a general agreement that within this context technology is given a broad definition to include not just hardware such as boilers or turbines but also plans, patents, designs, trademarks and copyrights collectively known as intellectual property rights (Davidson, 1991; Ketilsson, 1991; Robinson, 1991, 1988; Siddiqi, 1991; Tolba, 1991; Franklin, 1990; Rosenburg and Frischtak, 1985).Training in technological procedures as well as planning methods (e.g. environmental impact assessment) may also be considered as technology (Biswas and Geping, 1987).

There is not the same degree of consensus about what is meant by the transfer of technology. The difference of opinion regarding what constitutes technology transfer often reflects the goals which it is hoped the transfer of technology will achieve. Some have suggested that technology transfer is a process – a series of activities and responses that, over time, facilitate the transfer of a specific technology from one area to another. This process approach may be seen in the context of business development as part of industrialization and economic growth (Goldemberg and Monaco, 1991; Ketilsson, 1991) or as an approach to technological change (Figure 9.1) reflecting a desire for an improved quality of life (de Oliveira, 1991; Gomulka, 1990).

Others see profit motive on the side of the sellers of technology as the driving force behind technology transfer. In this instance, technology transfer is considered a euphemism for technology trade. This view has been summarized as: 'You pay, you get; you don't pay, you don't get' (Galal, 1991). Indeed, both those involved in broader discussions of technology transfer and those in a position to offer technology for sale seem to agree that market principles are at odds with simply giving technology away (Coleman, 1991; Tolba, 1991).

Technology transfer takes place in a number of forms. Figure 9.2 illustrates some of the basic variations ranging from adaptive transfer, the supply and immediate adaptation of technological information to local conditions prior to production, to pseudo-transfer, the use of imported technology for production in a foreign-owned plant, with no real transfer of skills or hardware (e.g. local workers are employed for non-skilled positions only) (Parthasarathi, 1988).

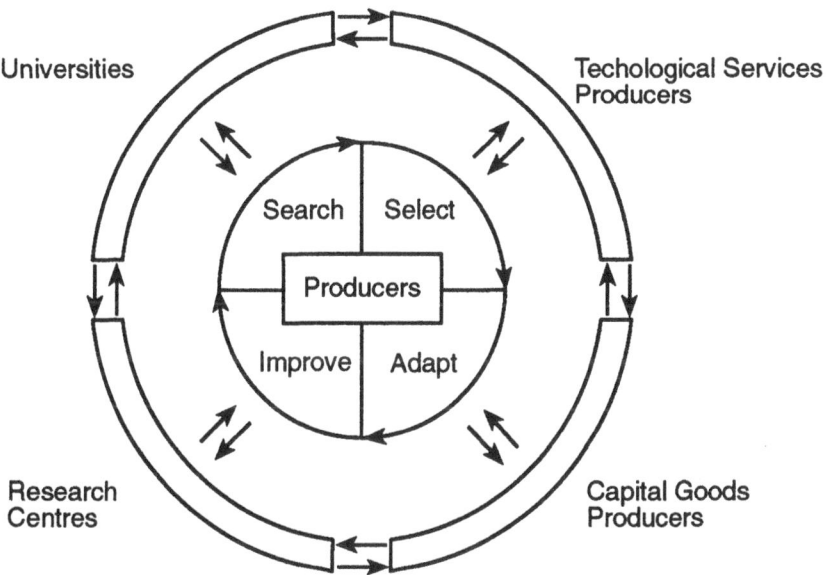

Universities

Techological Services
Producers

Search | Select

Producers

Improve | Adapt

Research
Centres

Capital Goods
Producers

Figure 9.1 *The technological array and the technological change
process*

(Source: de Oliveira, 1991)

Controversy surrounding technology transfer

As was mentioned earlier, a number of technologies are currently available
which can aid in the elimination or prevention of negative environmental
effects during low-grade fuel use. There is, however, a debate raging over
the most suitable technologies for alleviating environmental problems
where no previous environmental protection measures have been
implemented.

New and expensive versus older and cheaper

One area of controversy in this debate centres on the introduction of the 'best
available technology' for environmental control versus the application of
often simpler and less costly technologies. Questions raised in this regard
often echo discussions of large-scale versus small-scale development and
environmental quality which have been ongoing in earnest since the early
1970s.

A Adaptive transfer[1]

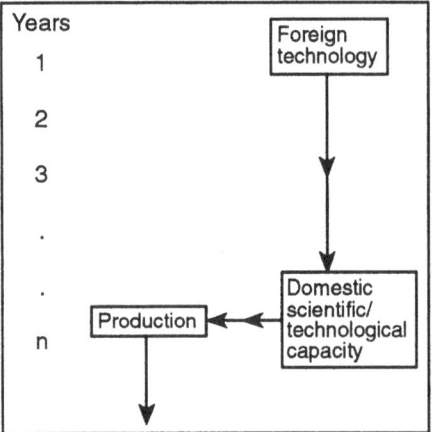

[1] Foreign technology supplied in Year 1 is adapted by a domestic technological and scientific capacity before going into production in Year Z

B Full transfer[2]

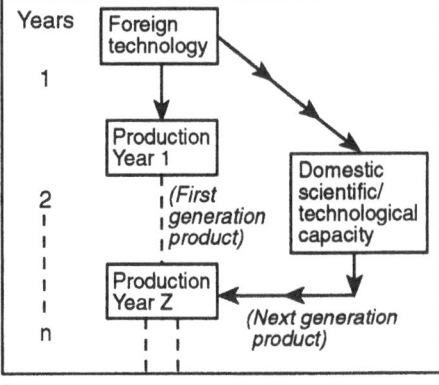

[2] Technology is purchased in Year 1, is simultaneously used in production and made the subject of domestic R & D and engineering design. In Year 'n' when the technology has to be renovated/upgraded, the domestic technological and scientific capacity will be able to deliver the renovated or so-called next generation technology.

C Full adaptive transfer[3]

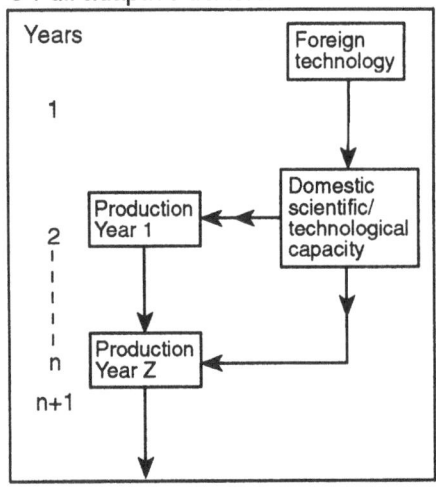

[3] A combination of A and B

D Pseudo-transfer[4]

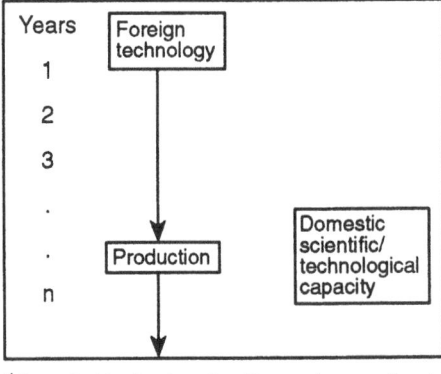

[4] Imported technology functions only as an input into production. Such an 'encapsulated form' of use of imported technology does not constitute a transfer at all. Its spillover effects are minimal; domestic skilled personnel perform a passive role.

Figure 9.2 *Forms of technology transfer and the development of domestic, scientific and technological capacity*

(Source: Parthasarathi, 1988)

Although the following quote refers to renewable sources of energy, the sentiment expressed is also relevant to technology transfer and low-grade fuels:

> Renewable energy projects should always be based upon adequately proven technologies. Rural areas of the developing world should not be seen as test beds for new technologies but rather areas of acute need which deserve to be helped in the most effective manner possible (Foley, 1991).

The PFBC system is a complex and expensive technology which some would offer for sale while others would argue is still in the experimental stage (Branwell, 1991). PFBC has shown excellent results for efficiency and clean burning of low-grade fuels (Patterson, 1990). The cost and expertise required to operate a PFBC system must be weighed against future technological innovations and the technical options available to deal with the environmental impacts which PFBC would be employed to control.

There is also a belief that it is important to 'do the simple things first' (Couch, 1992), such as assessing the specific problems of a system and determining where appropriate improvements may be made rather than replacing an entire plant. For the problem of increased sulphur emissions stemming from the combustion of high-sulphur lignite, relatively inexpensive technologies such as baghouses and ESPs could be installed to reduce the potential of a decrease in air quality (Arany et al., 1992). These technologies would not be useful in maximizing the efficiency of energy conversion, but they could make a significant contribution towards environmental protection.

The example of FGD is relevant here since over the last 20 years an enhanced understanding of the system chemistry of scrubbing has dramatically reduced the incidence of plugging, corrosion and erosion as well as land requirements associated with earlier models. Leading researchers in the field predict that innovations such as the application of spray drying to high-sulphur coal and the development of smaller components for use at sites where space is at a premium will not only result in better SO_2 control but may cost 20–50 per cent less to build and operate than the FGD systems on the market today (Torrens, 1992). This is an illustration of the difficulty that may arise when attempting to determine not only the most suitable technology for a given situation but also the most appropriate time for technology transfer to occur.

Another area of debate is whether developing countries and European economies in transition should be sold or given technologies which are no longer the technology of choice for new projects in developed countries where more effective technologies for accomplishing a task now exist. Passing on older technologies may be seen by some as off-loading unwanted goods onto a softer market or as a more sinister attempt to keep poorer countries reliant on systems and approaches which are 'second-best'.

The social impacts of technology

Most of the discussion surrounding technology transfer focuses on identifying and improving upon the market mechanisms which will facilitate the transfer of environmentally sound technology. There are, however, other facets of the debate which should not be overlooked. Of paramount importance is the recognition that there are values inherent in the design and use of any technology. These values are often culturally specific in that, within a given society, a certain technology will be seen as the one best way of solving a problem (Ellul, 1981). For this reason, there are those who argue that the social impacts associated with the transfer of technology must be considered side by side with the economic feasibility and desired degree of environmental protection.

Indigenous technology

The best technology is indigenous technology – there is nothing to replace it (Addo, 1991).

Criticisms have been raised regarding 'donor-driven' technological transfer, that is, technological aid given primarily for the benefit of increased sales in certain industries of the donor country (Addo, 1991; Davidson, 1991; Varadarajan, 1991; Adams, 1986; Goulet, 1977; Ladière, 1977). This is often one aspect of what is known as 'tied aid' which obliges the receiving country to accept aid in the form of goods and services as determined by the donor country. This has sometimes resulted in expensive technology being exported to a less-developed country where a genuine need has not been identified and the adequate training and build-up of infrastructure has not taken place. In turn, this may lead to the eventual abandonment of the technology as it begins to break down and the parts and expertise and perhaps even desire to repair the equipment are not available (Butera, 1989; Adams, 1986).

A sensitivity to local conditions is essential for the successful transfer of technology. As one observer has noted, the best way to achieve this is to 'plan with the people not for the people' (El Gizouli, 1991). Failure to adopt technology to the local situation can result in a number of problems, including:

- a concentration of income in a very small percentage of the population;

- underutilization of abundant local human and natural resources;
- heavy use of and reliance upon imported inputs; and
- concentration of scarce domestic and imported technological capabilities in the modern sector of society increasing the disparity between this sector and the traditional and rural sectors (de Oliveira, 1991; Stewart, 1977).

If the goals of technology transfer include technological and industrial maturity in a specific field, then the importation of expertise and technological hardware is important (Jhirad, 1990). Some would argue, however, that a primary goal is the stimulation of indigenous technological innovation in less-developed countries (Addo, 1991; Davidson, 1991). This would require the transfer of resources for research and development of technology rather than the transfer of technology itself. This may be at odds with those acting from the perspective of technology transfer as technology trade.

Effective technology transfer

Over the last three decades there has been a growing understanding of the approaches and barriers that encourage or prevent the successful transfer of technology. Successful technology transfer requires a thorough awareness of the context within which the transfer is taking place. Foremost, this necessitates an explicit understanding and agreement by all parties involved as to the desired goals and objectives of the transfer.

Experience has shown that technologies are adapted most successfully where the demand for technology has originated from the receiver, often in response to a genuine local need. Sound technology transfer requires adequate resources for the development of related infrastructure both technically (e.g. roads and communication systems, education and training) and politically (e.g. development and enforcement of necessary policy). All these factors are most likely to occur within an environment of political stability and mutual respect and understanding between or among the individuals and cultures involved (Hoskins, 1992; Oka and Jovanovic, 1992; Davidson, 1991; Goldemberg and Monaco, 1991; Contractor, 1980).

Methods and mechanisms of technology transfer

There is a wide range of methods used to transfer technology, many of them involving sales of hardware and various licensing agreements. A list of the major transfer mechanisms is given in Table 9.1. The application of these mechanisms is discussed in the broader framework of approaches to technology transfer in this chapter.

Table 9.1 *Methods and mechanisms of technology transfer*

- sales contracts for capital goods
- licensing
- consultancy
- technical training
- foreign investment in local enterprise
- minor innovation in existing plant
- technical education through scientific journals and patent descriptions
- transfer of person or institution
- visits to existing facilities
- policy initiatives
- joint research and development programmes
- establishment of co-operative effort outside sales initiatives
- exchange programmes
- international conferences
- expansion of multinational corporations

Constraints to technology transfer

The greatest constraints to the transfer of technology are often related to the lack of a specific commodity or necessary condition. Most commonly, these include a scarcity of funds or a lack of will or understanding on the part of either the receiver or the supplier. The most frequently cited constraints to technology transfer are given in Table 9.2.

Table 9.2 *Constraints to technology transfer*

- conflicting views as to the goals of technology transfer
- scarcity of financial resources
- lack of appropriate training/shortage of skilled labour
- inadequate infrastructure
- no genuine need for the technology
- conflict of commercial interests
- imposition of external values on the domestic situation
- competition with existing needs (e.g. industrial growth versus health and education)
- lack of political will

Approaches to environmentally sound technology transfer

The transfer of technology can take place within a number of contexts and using a variety of approaches. The activities described below indicate some of the current major trends in this area.

Business transaction between companies

Although technology transfer is often referred to as a transfer between countries, the most frequent method of transferring technology is from one business organization to another (Kristoferson, 1991). This may involve the use of sophisticated sales techniques and detailed contracts. The most successful of these endeavours include follow-up training and maintenance contracts as well as exchanges between individuals doing similar tasks relating to the technology (Hoskins, 1992).

Environmental considerations will frequently be taken into account if the companies selling the technology abide by the environmental regulations and legislation of their home country. The International Chamber of Commerce (ICC) has also recently released guidelines for conducting business in an environmentally sustainable manner (ICC, 1991).

Aid agency funding

International aid and banking agencies provide a substantial funding for development purposes which usually includes an element of technology transfer. Assessing the potential impact of the transfer of technology will aid in ensuring that the technology which is introduced is adapted and becomes integrated in existing local conditions.

Adequate funding for this type of assessment should be included in any project financing. It may form part of an environmental impact assessment, a requirement which is becoming mandatory for project approval by many aid agencies. For example, the World Bank now requires environmental screening of all proposed projects. Projects which fall under 'Category A', those which may have 'diverse and significant' environmental impacts, must undergo an environmental assessment including the identification of appropriate mitigation measures before funding will occur.

Enforcement of global standards

The enforcement of global standards for environmental quality could create a 'green' consumer market in which an increased volume of sales could lower the price of environmental technology. One method for increasing global environmental standards is through provision in trade agreements linking trade with proven environmental protection. This approach would be supported by improved global communications and transport, the existence of international professional organizations and the growth of transnational companies.

Real costing

In conventional neoclassical economics the cost of environmental clean-up has seldom been included when determining the financial viability of a project. The real cost should reflect clean-up costs as well as the cost of depleting a natural non-renewable resource such as lignite or oil shale to help minimize the negative effects of a project on the environment (Daly and Cobb, 1989). Within this context, environmentally safe technology often becomes more economically feasible than technology which does not include an environmental protection component.

Trust funds

Companies which design and manufacture environmental technology are in the business of making a profit and so are reluctant to cut prices or give away technologies to poorer countries. One suggestion for circumventing this problem is the development of trust funds to raise money to transfer environmentally appropriate technology to developing countries and European economies in transition.

References

ADAMS, P (1986) *In the Name of Progress: The Underside of Foreign Aid.* Probe International, Toronto.

ADDO, K T (1991) *Personal Communication.* Ministry of Energy, Ghana.

ARANY, F C, BARTOL, D A, COURCOULAS, J H and KEANE, G J (1992) 'Partnerships for upgrading coal-fired power plants'. *Conference on Clean and Efficient Use of Coal: The New Era for Low-rank Coal.* IEA/OECD, Paris.

BISWAS, A K and GEPING, Q (1987) *Environmental Impact Assessment for Developing Countries.* Tycooly International, London.

BRANWELL, A (1991) 'Perspectives on energy in Eastern Europe'. *Conference on Problems of Energy Conservation Technologies and Environmental Protection.* Varna, Bulgaria.

BUTERA, F M (1989) *Renewable Energy Sources in Developing Countries: Successes and Failures in Technology Transfer and Diffusion.* ENEA, Rome.

COLEMAN, J R (1991) 'Motives and mechanisms for generating and transplanting new technology: the role of government and business'. *Conference on Technology Transfer and Global Environment: Motives and Mechanisms.* RIIA/WRI, London.

CONTRACTOR, F J (1980) 'The composition and licensing fees and arrangements as a function of economic development of technology recipient nations'. *Journal of International Business Studies* **11**: 48.

COUCH, G R (1992) 'Low-rank coal in Eastern Europe: opportunities and constraints'. *Conference on Clean and Efficient Use of Energy. The New Era for Low-rank Coal.* IEA/OECD, Paris.

DALY, H E and COBB, J B (1989) *For the Common Good.* Beacon Press, Boston.

DAVIDSON, O R (1991) 'Indigenous capacity in R + D: possible strategies of tapping local technical skills and capital'. *International Symposium on Environmentally Sound Energy Technologies and Their Transfer to Developing Countries and European Economies in Transition.* Milan.

DE OLIVEIRA, A (1991) 'Energy technology policy in developing countries'. *Science and Public Policy* **18**: 156–164.

EL GIZOULI, A (1991) *Personal Communication.* National Energy Administration, Sudan.

ELLUL, J (1981) *Perspectives on Our Age* (ed by W Vanderburg). Canadian Broadcasting Corporation, Toronto.

FOLEY, G (1991) *Energy Assitance Revisited: A Discussion Paper.* SEI, Stockholm.

FRANKLIN, U (1990). *The Real World of Technology.* Canadian Broadcasting Corporation, Montreal.

GALAL, E E (1991) *Personal Communication.* National Society for Technology and Economic Development, Cairo.

GOLDEMBERG, J and MONACO, L C (1991) 'Brazil: transfer and adaptation of environmentally sound energy technologies'. *International Symposium on Environmentally Sound Technologies and Their Transfer to Developing Countries and European Economies in Transition.* Milan.

GOMULKA, S (1990) *The Theory of Technological Change and Economic Growth.* Routledge, London.

GOULET, D (1977) *The Uncertain Promise.* International Documentation, New York.

HOSKINS, E W (1992) 'Short-term and long-term actions to increase technological co-operation with Central and Eastern European countries'. *Conference on the Clean and Efficient Use of Coal: A New Era for Low-rank Coal.* IEA/OECD, Paris.

INTERNATIONAL CHAMBER OF COMMERCE (1991) *Second World Industry Conference on Environmental Management.* ICC, Norway.

JHIRAD, D (1990) 'Power sector innovation in developing countries: implementing multifaceted solutions'. *Annual Review of Energy* **15**: 365–398.

KETILSSON, O (1991) 'Technology co-operation for sustainable development'. *International Symposium on Environmentally Sound Technologies and Their Transfer to Developing Countries and European Economies in Transition.* Milan.

KRISTOFERSON, L (1991) 'Comments: subsession B2.1'. *International Symposium on Environmentally Sound Technologies and Their Transfer to Developing Countries and European Economies in Transition.* Milan.

LADIERE, P (1977) *The Challenge Presented to Cultures by Science and Technology.* OECD, Paris.

OKA, S and JOVANOVIC, L (1992) 'Clean coal technology transfer and co-operation'. *Conference on the Clean and Efficient Use of Coal: A New Era for Low-rank Coal.* IEA/OECD, Paris.

PARTHASARATHI, A (1988) 'Acquisition and development of technology: the Indian experience'. *Intersciencia* 13: 241–251.

PATTERSON, W C (1990) *Coal Use Technology in a Changing Environment.* Financial Times Business Information, London.

ROSENBURG, G and FRISCHTAK, C (1985) *International Technology Transfer.* Praeger, New York.

ROBINSON, R D (1991) *The International Communication of Technology: A Book of Readings.* Taylor and Francis, London.

ROBINSON, R D (1988) *The International Transfer of Technology.* Ballinger Publishing, Cambridge.

SIDDIQI, T (1991) 'Technology transfer: trade, market and intellectual property issues'. *Conference on Technology Transfer and the Global Environment: Motives and Mechanisms.* RIIA/WRI, London.

STEWART, F (1977) *Technology and Underdevelopment.* Macmillan, New York.

TOLBA, M (1991) 'Facilitating technology transfer in the interests of the global environment'. *Conference on Technology Transfer and the Global Environment: Motives and Mechanisms.* RIIA/ WRI, London.

TORRENS, I M (1992) 'Fossil power plant control technologies: status, performance, cost'. *Conference on Clean and Efficient Use of Coal: The New Era for Low-rank Coal.* IEA/OECD, Paris.

VARADARAJAN, J (1991) 'Technology dissemination in developing countries: process and recipient needs'. *Conference on Technology Transfer and Global Environment: Motives and Mechanisms.* RIIA/WRI, London.

Costs considerations and low-grade fuel use

The use of low-grade fuels depends not only on their availability and supply security but also on the technological feasibility of procurement (mining or harvesting), transport, storage and the extent of waste generation on use, potentially to the atmosphere, water courses and land. Each stage of the fuel cycle has an environmental (or external) cost, and these costs should be internalized to the operation. They must be added to the other costs of planning, development, exploitation and use of the fuel, which are the industry or internal costs.

There are well-established procedures for obtaining financial estimates for planning, development activities, exploitation and operations, and also the direct costs of waste treatment and disposal. The extent of the true environmental costs may, however, be much greater and more difficult to estimate (Laikin et al., 1991; OECD, 1977). One sub-set of environmental costs, often treated separately, is the health costs resulting from pollutant emissions. These can be dealt with in a more straightforward fashion, as it is possible to evolve some cost functions applicable to human morbidity or death. For example, lost days at work or loss of earnings due to death can be costed, and sometimes are for compensation purposes.

Total environmental costs are more difficult to deal with. It is not satisfactory merely to estimate these as costs involved in fitting 'end-of-pipe' control technologies. These may not be capable of reducing the problem to an environmentally acceptable level. It may be necessary to completely redesign a production process to a low-waste level so that environmental goals are met by cleaner production practices.

The costs associated with the considerations above apply to many solid fuels without any clear pro rata variation with the calorific value of the fuel. Thus, for low-grade fuels there will not necessarily be any overall reduction of either industry (internal) or environmental (external) costs implied in their use. It is unlikely, therefore, that any cost advantage will be a determining factor in the preferential choice of low-grade fuels; indeed, the reverse may be the case. It is likely that other considerations (availability, domestic supply, linkage to other issues such as waste disposal) will play a large part in determining whether or not, or to what extent, low-grade fuels will be utilized.

Industry or internal costs

The development of any source of fuel for exploitation involves costs. These costs will be incurred long before any benefits arise from the exploitation of the fuel. The costs will arise in the first instance from the initial construction works and the provision of infrastructural facilities such as roads, housing and other service requirements. The mine establishment itself will involve costs, for construction of the mine, costs for related facilities (such as railways and loading facilities) and then costs associated with the operation of the mine. Capital costs arising from construction of the mine are linked to capacity factors and are not usually very easily recouped until well into the production cycle; the balance of operating costs and accruing revenue are closely linked to levels of production.

The process of beginning the exploitation of any low-grade fuel resource should be preceded by obtaining estimates of costs as accurate as possible and the production of a financial spreadsheet. This should summarize the total costs and expected revenues and should be realistically scheduled. A series of revenue estimates should be included. Under certain arrangements the distribution of costs and revenues will also be of great importance.

The financial aspects of the construction and operation of a mine to obtain low-grade fuel, or an associated facility, are not unrelated to consideration of the costs stemming from potential environmental and health impacts. These costs, if internalized in a satisfactory manner, will be incurred during both construction and operation. However, it will be during operation that the longer-term costs will arise.

External costs

Conventional costing usually focuses on identifying the internal costs of an energy project leaving aside, for example, longer-term issues such as waste management and potential disruption to air and water quality and land use. Environmental costs are generally considered as externalities which do not require costing. However, there are real costs associated with a deterioration in environmental quality and, increasingly, proponents of energy projects are being encouraged to take into account the total environmental costs of the project, that is, to include the costs of managing or dealing with externalities.

As mentioned earlier, the costs for including environmental protection, in the broadest sense, may be much greater than conventional costing exercises reveal and also be more difficult to determine (Laikin et al., 1991; OECD, 1977). For example, small-scale chronic emissions to the air may cause only small immediate changes in ecosystem structure and function, possibly in areas many kilometres from the emissions source. Over time, however, the initially slight changes may result in sustained nutrient loss, reduced productivity, changes in species composition and other alterations for which no cost has been attributed to the original emitting source. Similar unpredictable long-term effects may also be envisaged for effluent disposal to lakes and streams. There are, in fact, many potential pollutants, each of which has its own set of hidden cost implications.

As the above example illustrates, it is not sufficient merely to estimate costs involved in fitting end-of-pipe control technologies. Particularly in the long run, these types of solutions may not achieve a cost-effective environmental protection. True costing may indicate the necessity to redesign a production process (such as making use of low-waste technology) in order to meet environmental goals through cleaner production practices.

Environmental loads

Attempts have been made to set the level of pollution that can be tolerated by trying to establish the assimilative capacity of an ecosystem (or related measure such as absorptive capacity, receiving capacity, environmental capacity or critical load). These concepts have been developed and expounded by Pradvic (1987), Nilsson (1986), Cairns (1977) and Velz (1976). They involve the ability of an ecosystem to receive a substance at a rate (or in an amount) that will not degrade its structural or functional attributes. The financial implications that ensue from not allowing emissions to occur at levels or rates that would result in exceeding this capacity are often not only great but also far-reaching. Costs may also be high if an attempt is made to achieve abatement through applying the best available technology (BAT) irrespective of the assimilative capacity of an impacted ecosystem.

Health costs

Health risks resulting from pollutant emissions give rise to external costs. In general, health costs are easier to calculate than other environmental costs since estimates can be made of the financial burden arising due to human morbidity and death. In the field of both occupational and public health, exposure levels are set at a level that attempts to make the risk of sickness, injury or death very low. Exposure levels even take into account specific sensitive groups (infants, the elderly, diabetics or those with chronic illnesses). Many studies and assessments have been undertaken (Hamilton, 1984, 1979) to a point where comparative risk assessments may be carried out. Epidemiological studies are able to relate changes in environmental quality to health detriment, and costs can be evolved as a measure of this.

Costs of control technology

The cost of the installation of control technologies in new plants, or the retrofitting of old ones, to ensure high environmental performance can be determined relatively easily. The operation and maintenance costs of such technologies can also be estimated. These costs have to be added to the production costs and, in some way, recovered.

In attempting comparisons from country to country, problems of differing raw materials costs, labour costs, interest rates, inflation rates and other considerations need to be addressed. For these reasons it may be more convenient to resort to the expression of costs in terms of, for example, the energy penalty incurred per unit of pollutant removed. This approach is considered below.

Technology cost estimates

The cost estimates of applying a pollution control technology depend, crucially, on a number of factors, but since each situation is unique it is impossible to arrive at universally applicable cost estimates. The factors that determine pollution control costs include:

- fuel characteristics;
- conversion route characteristics;
- pollutant to be controlled;
- size of facility;
- control technology employed;
- rate or amount of fuel use;
- labour costs;
- raw materials costs;
- prevailing financial conditions;
- existing environmental conditions.

Table 10.1 presents the main technologies for the control of SO_2 and NO_x emissions from the use of lignite, peat, briquettes and oil shale. For the control of particulate matter bag filters, ESPs or cyclonic devices would be employed, depending on the size of the facility.

Table 10.1 *Emission control technologies for SO$_2$ and NO$_x$ employed with a number of low-grade fuels*

Fuel	SO$_2$ emission control			NO$_x$ emission control	
	Sorbent injection	FGD	FBC	Combustion modification	SCR
Lignite:					
existing	x	x	–	x	x
new plants	x	x	x	x	x
Peat:					
existing	x	x	–	x	x
new plants	x	x	x	x	x
Briquettes:					
existing	x	x	–	x	x
new plants	x	x	x	x	x
Oil shale:					
existing	–	x	–	x	x
new plants	–	x	–	x	x

(Source: EPRI, 1985)

Figure 10.1 indicates the range of costs, in relation to the extent of pollution emission reduction, that might be encountered. The construction of such cost-control relationships is possible for countries, a region or even for an individual plant. Usually this will be done for a single pollutant or a group of closely related ones.

Cost estimation methodology

Specific emission control technology applications can be costed (Rentz, Remmers and Plinke, 1988; UN-ECE, 1988). Where these are applied, it is usual to attribute costs on an annual basis. Thus investment and operational costs are apportioned by taking into account the estimated life-time of technological application. This involves calculation of capital cost, discount rate and economic lifetime estimates. Operating and maintenance costs can then be added on an annual basis.

Once abatement costs for removing individual pollutants have been determined, it is possible to calculate the cost per unit of pollutant removed. By combining the costs for various relevant technologies, a complete marginal cost curve may be obtained. When considering the costs of

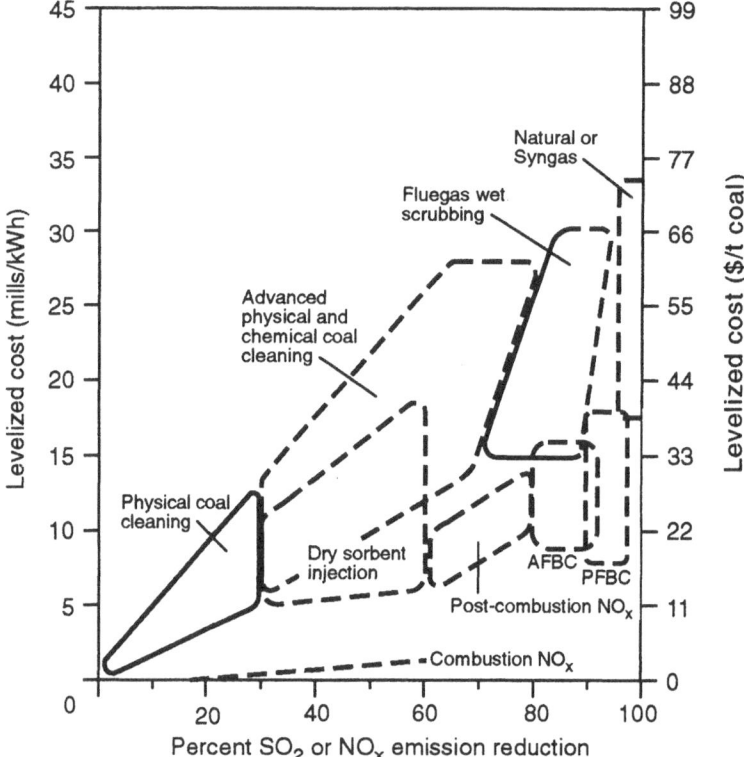

Figure 10.1 *Comparison of cost and efficiency for emission control*

(Source: EPRI,1985)

removing an individual pollutant using different technologies, the relative per unit cost of pollutant control becomes apparent. Figures 10.2 and 10.3 show examples of marginal cost curves for two countries for sulphur abatement. It should be noted that they involve fuels other than low-grade fuels. They do illustrate the fact that single cost curves for use from country to country are not appropriate.

Other methodological approaches

In order to increase the ease of comparison, it has been suggested that costs of emission reductions might be expressed (for electricity-generating plants for example) as MWh per unit of pollution abated. Table 10.2 shows the results of such an approach for NO_x and SO_2 emissions reduction in coal-fired boilers.

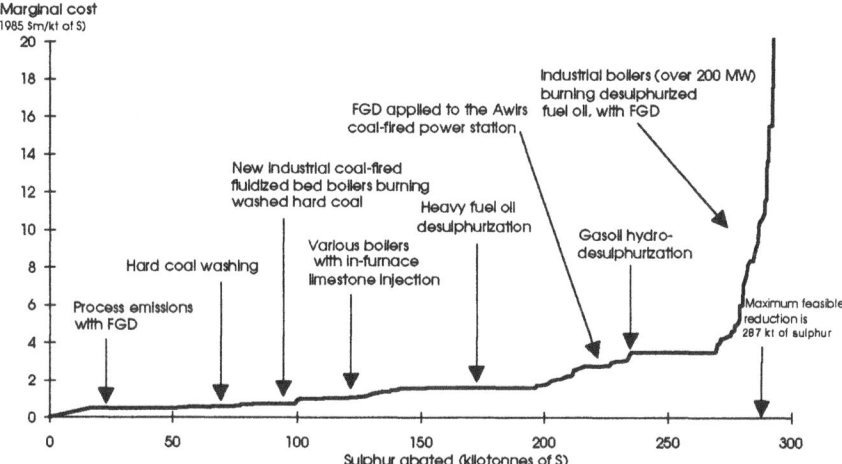

Figure 10.2 *Marginal cost curve of sulphur emissions abatement in Belgium for the year 2000, with a number of the abatement measures indicated*

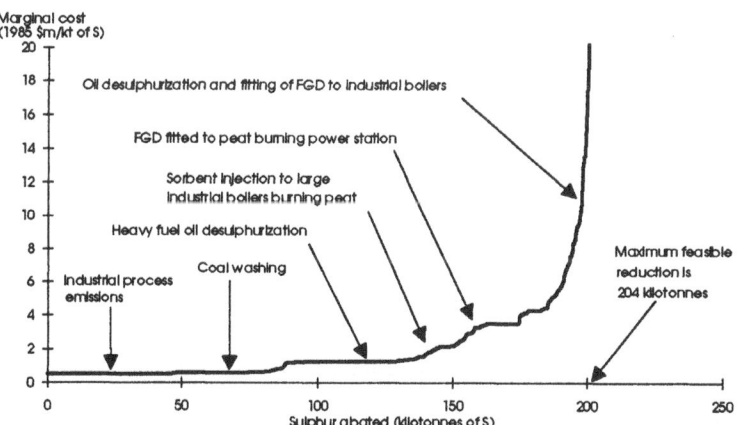

Figure 10.3 *Marginal cost curve of sulphur emissions abatement in Finland for the year 2000, with some of the abatement measures indicated*

Table 10.2 *Cost of emission reduction in terms of electric power generated*

Method of control	Removal (mg·m^{-3}*)	Electric power cost of pollutant reduction (MWh·t$^{-1)}$
NO$_x$:		
Combustion modification	350–500	5 – 7
SCR	200–240	50–70
SO$_2$:		
Sorbent injection (low S)	1100–1500	7–15
FGD (high S)	2500–3900	15–23

* As NO$_x$ or SO$_2$: initial pollutant concentration not equal for the different technologies applied

Pollution abatement options

End-of-pipe technological abatement solutions may not always result in the required emission reductions. A range of 'before', or 'during' combustion options (of which sorbent injection and FBC are examples) may provide enhanced abatement opportunities. The possibilities of using advanced technologies such as combined-cycle gas turbines with low calorific value fuels have not yet been widely investigated. In addition, technologies able to contribute to the control, simultaneously, of more than one pollutant might have cost-effectiveness advantages, but the low-cost margins implied in the use of low-grade fuels mean these have not been explored.

References

CAIRNS, J Jr (1977) 'Aquatic ecosystem assimilative capacity'. *Fisheries* **2**: 5–7 and 24.

EPRI (1985) *Briefing to US Senate Committee on Acid Rain*. EPRI, Palo Alto (unpublished).

HAMILTON, L D (1984) 'Health and environmental risks of energy systems'. In *Risks and Benefits of Energy Systems*. IAEA, Vienna.

HAMILTON, L D (1979) *Comparing the Health and Environmental Hazards of Different Energy Systems*. Brookhaven National Laboratory, Upton, NY.

LAIKIN, R, SCHÄRER, B, REMMERS, J, PLINKE, E, KLAASSEN, G and HAASIS, H D (1991) 'Control costs of SO$_2$ and NO$_x$ emissions'. In *Acid Depositions in Europe* (ed by M J Chadwick and M Hutton) SEI, Stockholm.

NILSSON, J (ed) (1986) *Critical Loads for Nitrogen and Sulphur: Miljorapport 1986* Vol 11. Nordic Council of Ministers, Copenhagen.

OECD (1977) *Energy and Cleaner Air: Costs of Reducing Emissions*. OECD, Paris.

PRADVIC, V (1987) 'Environmental capacity: an approach for prevention of water pollution'. *Water Quality Bulletin* **12**: 137–40 and 166.

RENTZ, O, REMMERS, J and PLINKE, E (eds) (1988) *ECE Workshop: Emission Control Costs. Methodology and Example Cases*. IIP, Karlsruhe.

UN-ECE (1988) *Draft Guidelines for the Cost of Emission Control Activities*. UN-ECE, Geneva.

VELZ, C J (1976) 'Stream analysis: forecasting waste assimilative capacity'. In *Water Resource and Pollution Control* (ed by H W Gehm and J I Bregman). Van Nostrand Reinhold, New York.

Integrated planning and environmental management of low-grade fuels

Energy planning, like all forms of planning, is a process. One of the major aims of this process is to ensure that adequate primary sources of energy are available to meet the need for energy services such as transport, heat and power. But energy cannot be planned in isolation. It has linkages with the rest of the economy. Indeed, energy provision is often seen as the key to economic development. But the energy sector also has complex interlinkages between various subsectors within it. There may be conflicts within the sector on both demand and supply sides. One object of energy planning is therefore to optimize the use of the various fuels available in the range of economic sectors that constitute the demand. Another objective is to allow and plan for the development of energy resources in a balanced way, over time. Yet another aim is increasingly becoming to encourage energy efficiency and promote energy conservation. In recent years also there has been an increased awareness of the necessity of including environmental protection as one of the major factors constraining the pattern of energy use.

All of these energy-planning considerations are relevant to low-grade fuel use. Previous chapters have stressed that the use of peat, oil shale, coal, wood and other fuels gives rise to the potential for environmental disruption and contamination (in different ways for different fuels) from activities such as mining, fuel transport and storage, upgrading and processing, and combustion. These considerations emphasize the need for an integrated approach to the planning of their exploitation and use, and the management of the complete fuel cycle.

Integrated energy planning

Munasinghe (1990) regards the broad rationale underlying integrated energy planning (IEP), and integrated national energy planning (INEP), specifically as the need for an energy master plan (EMP). This is required to 'make the best use of energy resources, promote socio-economic development and improve the welfare and quality of life of citizens'. Within these aims should be seen the necessity to maintain (and improve) the environmental quality base as one of the features by which the 'best-use' of energy resources are obtained and the 'welfare and quality of life' of the population are enhanced.

An INEP framework is depicted in Figure 11.1. The framework will aid in the development of coherent policies that attempt to meet what are interrelated but often conflicting national objectives (Munasinghe, 1990).

Features requiring integration in IEP occur at a number of levels. An energy plan must be an integral part of the overall economic plan for the country. It should take into account the plans of the different energy subsectors and should also integrate the individual components of each energy subsector. Access to information is essential for assembling database for specific use in the IEP (Figure 11.2).

In establishing this energy database, which will, ideally, be continually updated, it is necessary to have in mind the policy objectives. It is possible to make use of certain 'formalized' data-holding structures, usually taking the form of computer software packages. One that has been extensively used in developing countries is the long-range energy alternatives planning system (LEAP), an energy planning tool developed by the Stockholm Environment Institute (SEI-B, 1992).

Using existing data as a starting point, a number of economic scenarios (often referred to as economic growth scenarios) are developed, giving consideration to major economic sectors such as industry, agriculture and forestry, transport, residential and commercial and others. From these sectors will emerge an overall energy demand scenario (or projection). A complementary step is to make an assessment of energy resources. This will entail identifying resources and reserves. In addition, the availability and price projections of international energy resources will need to be determined.

It is usual to proceed next to an evaluation of supply technologies. This is often carried out specifically with respect to fossil fuels, renewable resource technologies and the electricity supply system. However, there is no reason not to include other categories such as low-grade fuels. The criteria for evaluating technologies should be engineering performance; economic performance; environmental impacts; and technological modification and development. It is then necessary to balance supply and demand for any scenario generated, and energy flows need to be traced from source to end use.

The next step involves impact analyses. Impacts on the macroeconomic structure and growth rate, including a sector-by-sector assessment, impacts on inflation, consumption and the external trade and balance of payments situation are required. All of this needs to be related to a careful and realistic overall environmental impact analysis. To implement scenarios features an investment and financial plan will be required, closely linked to a supply and demand management strategy. This strategy will identify pricing and non-price options and is crucial in its environmental management

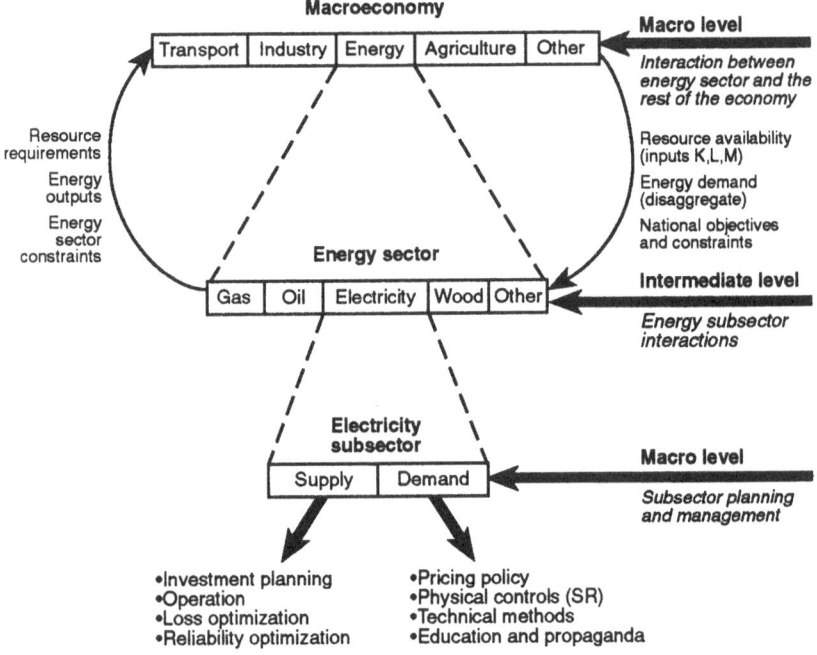

Figure 11.1 *Integrated national energy planning framework*

(Source: Munasinghe, 1990)

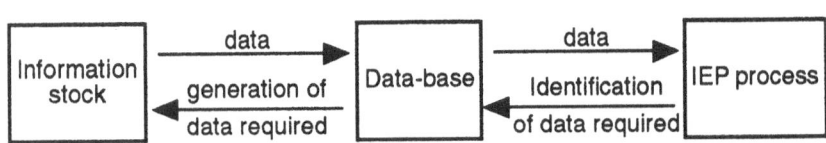

Figure 11.2 *The generation and identification of data for the IEP process*

(Source: APDC, 1985)

role. This IEP process is documented in detail by APDC (1985) and the overall process linkages are displayed in Figure 11.3.

Environmental impacts and integrated energy planning

In the past, during energy policy development, there has been a tendency to overlook the negative pressure placed on the environment due to energy use but, increasingly, this is changing. Climate change induced by CO_2 has heightened awareness, on a global scale, of the relationship between energy use and environmental problems. Although energy use is still considered a major factor in determining the level of economic activity, those responsible for energy planning are beginning to investigate approaches that will take into account the well-being of the environment in tandem with the exploitation of energy (Asian Development Bank, 1992; IAEA, 1991; OECD, 1991a; Danish Ministry of Energy, 1990).

Integrating the environment into energy-related decision making requires an understanding of the linkages within and between the environment and an energy system. This understanding goes hand in hand with the development of better methods for quantitatively identifying and assessing the aspects of energy use which have a negative effect on environmental quality. A discussion of energy sector indicators of environmental quality is presented in this chapter along with current approaches to building environmental protection into energy projects and plans. These approaches include environmental impact assessment, computer-generated models and restructuring institutional mechanisms. Conservation and energy efficiency, often a response to problems arising from the energy/environment relationship, are considered in Chapter 8. Cost factors governing some aspects of integrated energy/environmental planning are covered in Chapter 10.

Energy sector indicators of environmental quality

The development and measurement of indicators of the energy/environment relationship form an important component of IEP and will ultimately contribute to an enhanced comprehension of what is meant by sustainable development with respect to energy. Conventional indicators have focused on the amount of energy used as a measure for the potential of environmental degradation (OECD 1991a, 1991b). This reflects the accepted linkage between the quantity of energy used and economic growth and the belief that greater energy use equals greater industrial output. A more recent indicator of the relationship between energy use and the environment is the consideration of energy intensity.

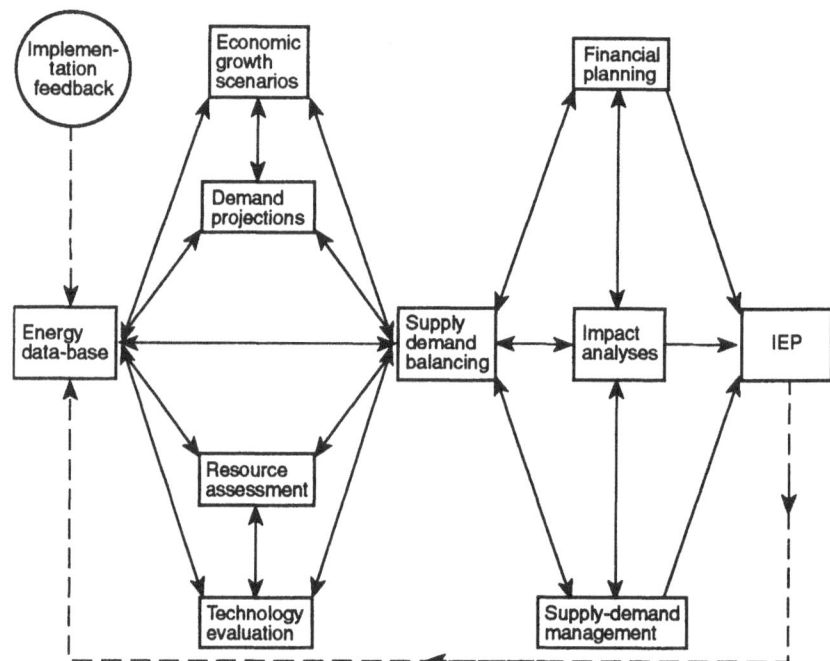

Figure 11.3 *The IEP process*

(Source: APDC, 1985)

Energy intensity

Energy intensity, in this context, refers to the amount of energy consumed per unit of gross domestic product (GDP). As Figure 11.4 indicates, in OECD countries conservation policies and the increased use of energy-efficient technology have contributed to a decrease in the quantity of energy required to produce a unit of GDP over the last 20 years.

The information presented in Figure 11.4 takes into account conventional primary sources of energy including nuclear, gas, oil and solid fuels. Energy intensity is an important environmental indicator for fuels such as oil shale, peat, wood, lignite and municipal and industrial waste as well, since these fuels as mined, harvested or collected have a low energy density. This creates the potential for greater environmental disruption than with

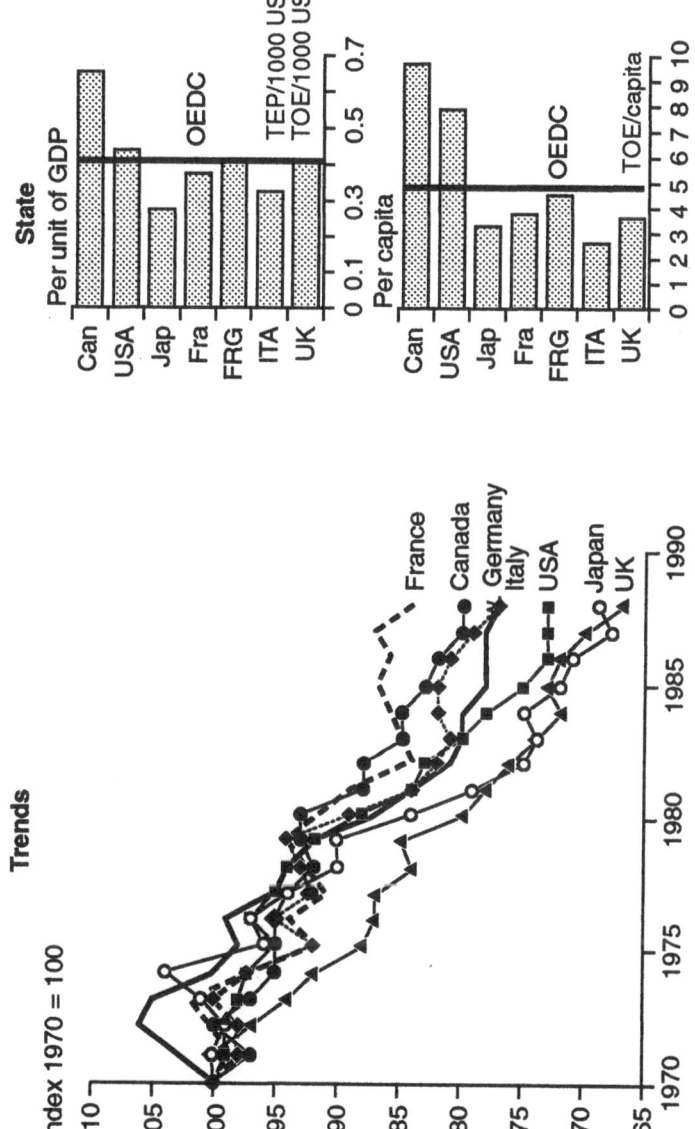

Figure 11.4 *Trends in energy consumption for selected OECD countries*

(Source: Munasinghe, 1990)

conventional fuels, particularly when low-grade fuels are used on a large scale in industry or for power production and emphasizes the importance of using energy-efficient technologies with low-grade fuels.

Environmental impact assessment

Environmental impact assessment (EIA) is the process of evaluating the likely consequences of a proposed action and can be carried out for projects, proposals, plans and programmes, thus providing decision-makers with an indication of the potential outcomes of their actions. Ideally, an EIA should be carried out during the pre-planning stage, when the possibility of project rejection remains open if an assessment indicates that the environmental costs of a proposal are too high. However, the effectiveness of EIA as an environmental management tool will always be limited by the degree to which the relevant authorities are willing or able to act on the study findings.

Conducting an EIA

The major steps involved in conducting an EIA include:

- definition of the undertaking including a clear statement of the goals of the proposed action (DESIGN);
- identification of the components of the environment which will be included in the study (e.g. water quality, air quality, economic costs and benefits, social considerations such as community structure) (SCOPING);
- a study of existing conditions (BASELINE);
- prediction of type and severity of potential environmental effects (PREDICTION);
- identification of measures to mitigate or eliminate potential negative environmental impacts (MITIGATION);
- summary of overall potential environmental impacts of the proposal (EVALUATION);
- design of a programme for monitoring the actual environmental impacts of the project (MONITORING).

At various stages throughout the process the public should be given an opportunity to review and comment on the EIA. The steps outlined above will generally result in a document which is known as an environmental impact statement.

The collaboration of a multidisciplinary team, representing, for example, hydrology, agriculture, sociology, botany, zoology, economics, anthropology, atmospheric science, hydrogeology, engineering and geology, offers the greatest likelihood of predicting the majority of potential impacts associated with a proposal. Once these impacts have been identified, the goal of integrating environmental considerations into energy planning will be furthered by implementing measures to control or eliminate undesirable effects.

Comparing alternatives

EIA can be useful for choosing among alternatives. There is almost always more than one method for achieving a particular development, and the choice of one alterative over another for expediency or, on economic grounds alone, may lead to future costly environmental problems.

Some of the greatest concerns regarding low-grade fuel use stem from the potential air and water pollution associated with combustion. The type and quantity of pollutant emissions can frequently be controlled by the choice of combustion technology. FBC is often more suitable than conventional boilers for low-grade fuels, since FBC can achieve more complete burning of fuels with a wider range of characteristics. Increased combustion efficiency can decrease pollutant emissions and reduce the amount of primary energy needed to meet demands.

Table 11.1 presents a comparison of different combustion methods used for low-grade fuel in Finland.

Modelling approaches

Computer models can be useful for highlighting key relationships within the energy/environmental system. Models are only as good, however, as the assumptions from which they arise and the data upon which they are based. While computer models are sometimes regarded as complicated, they are in reality a simplification of various relationships within the natural world. The constraints on obtaining data, such as uncertainty regarding national reserves of a fuel or the relationship between the emission of certain pollutants and human health, can lead to a situation where what *can* be modelled takes precedence over what *needs* to be modelled (Biswas, Khoshoo and Khosla, 1990). In addition, computer modelling relies almost exclusively on quantitative data, which makes it difficult to include the more human aspects of a proposal such as social, cultural and political factors. For example, it may be difficult to incorporate the affordability of a proposed strategy on low-income groups.

The usefulness and applicability of models revolve around both the data requirements: availability, suitability and quality (parametric considerations)

Table 11.1 A comparison of combustion methods[a,b] for solid fuels in Finland

	Pulverized combustion	Grate combustion	FBC bubbling	FBC circulating	Cyclone combustion	Gasification combustion
Capacity	50–720 MW$_{th}$	1–70 MW$_{th}$	3–70 MW$_{th}$	5–240 MW$_{th}$	10–40 MW$_{th}$	5–35 MW$_{th}$
Fuels	Coal	Wood waste	Peat	Peat waste	Wood waste	Wood
	Peat	Peat	Wood waste	Wood waste	Peat	Peat
	Wood waste	Coal	Coal	Coal		
		Sludge	Sludge	Sludge		
Boiler Efficiency	92%	88%	92%	92%	85%	99% (only for gasification)
Special Features	High potential for negative environmental impacts.	Used in district heating and industrial applications.	Low NO$_x$ concentration in flue gas. SO$_2$-removal also possible in furnace.		Very seldom used today; old-fashioned method.	Used in district heating and industrial applications.

Table 11.1 (continued)

Pulverized combustion	Grate combustion	FBC bubbling	circulating	Cyclone combustion	Gasification
Most frequently used in large-scale power production.	Limited poten-tial for multi-fuel use.	Good potential for multifuel use. Used in district heating and industrial applications.		Most common-ly used in industry.	

[a]Table does not include combined methods such as pulverized-grate firing, which are in use
[b]Includes subcategories such as mill combustion and drier pulverized combustion

(Source: Tuovinen, 1992)

and the relationships between components within the model: the linkages and feedback loops (the structure of the model). It may be possible to obtain data for a simply-structured model so that a 'data-rich' but rather transparent model is constructed. Another approach would be to emphasize the linkages, at a number of levels, so that the structure attempts to mirror inter-dependencies in the real world. A model of such complexity might not be able to obtain the necessary data sets, but its behaviour could be investigated to produce more 'metaphorical' indications that would enable insights concerning the real world to be obtained.

The above discussion points out the limitations of computer modelling for planning purposes, but in spite of certain weaknesses computer models can prove to be useful tools in the planning process as already indicated. A model may significantly add to the quantity and quality of information available to decision makers, and when used in conjunction with knowledge regarding local conditions and sound judgement resulting from experience, it can increase the likelihood of successfully integrating the environment into energy planning.

Computer models may be applied to further the goal of environmental management of energy in several ways. They may be used as planning tools for tasks such as tallying resources available to meet energy demand, to which an environmental component has been added (e.g. determining the quantities of potential pollutants released, using different fuels, under a variety of demand scenarios). LEAP is a series of user-friendly microcomputer programs for matching energy supplies to changing demands which has been used in 50 countries, including Kenya, Tanzania and Senegal, for energy planning (SEI-B, 1992).

The LEAP energy assessment and planning system has provision for linkage to an environmental database (EDB) that gives indications of waste streams (emissions to air and water, and solid waste production) resulting from particular energy demand and supply situations or scenarios. An example is shown of the interlinked spread sheets in Tables 11.2–11.5.

There are other types of energy use impact models. It is possible to link fuel use in various sectors by a range of technologies to the emissions of pollutants that result (e.g. SO_2 and NO_x emissions to the air). Meteorological models such as EMEP (Eliassen et al., 1988) may then be employed to relate emission sources to patterns of deposition following atmospheric transport. If certain environmental criteria are established (or indeed, other criteria), it is possible to model the optimum control strategy (based on cost or some other constraint) that would meet those criteria. An outline of one such modelling approach (SEI, 1991) is given in Figure 11.5.

Table 11.2 *National energy demand scenario by fuel and year for all sectors (million GJ)*

	1985	1990	1995	2000	2005
Animal wastes	1.51	1.70	1.86	2.03	2.18
Aviation gas	0.45	0.53	0.60	0.66	0.72
Biogas	0.20	0.36	0.57	0.84	1.17
Charcoal	23.27	32.67	41.53	56.09	70.27
Coal, bituminous	2.80	3.80	4.79	8.33	13.01
Coking gas	–	0.04	0.09	0.37	0.66
Commercial wood	6.54	8.83	11.13	15.06	18.99
Diesel/gas oil	21.07	27.30	33.53	42.51	51.48
Electricity	5.84	8.88	11.96	16.58	21.26
Ethanol	–	–	–	5.96	14.98
Firewood	210.52	251.91	292.70	341.35	388.16
Gasoline	25.27	32.29	37.47	44.00	47.42
Kerosene/jet fuel	12.35	16.65	20.90	26.24	31.60
LPT/bottled gas	0.96	1.46	1.89	2.60	3.24
Metallurgic coke	–	0.25	0.50	2.11	3.72
Methane	–	0.02	0.04	0.15	0.26
Natural gas	2.12	3.57	5.02	8.45	11.89
Petrochemicals	1.52	2.55	3.59	6.04	8.49
Producer gas	0.11	0.16	0.21	0.28	0.35
Residual/fuel oil	19.23	26.32	33.39	39.95	45.31
Solar	0.41	0.67	1.01	1.77	2.75
Vegetal wastes	8.45	10.50	12.57	15.53	18.49
Wind	0.17	0.21	0.27	0.33	0.39
TOTAL	342.77	430.67	515.59	637.22	756.78

(Source: Example of interlinked spreadsheet produced by LEAP)

Table 11.3 *Emissions to the atmosphere resulting from a national energy demand scenario (Table 11.2) in kg*

	1985	1990	1995	2000	2005
Carbon dioxide					
Non-biogenic	6.30	7.96	10.16	13.21	16.73 (10^9)
Biogenic	20.77	25.74	30.59	37.17	43.67 (10^9)
Carbon monoxide					
Total	1869.72	2312.81	2704.92	3308.12	3884.72 (10^6)
Hydrocarbons					
Total	265.63	335.88	404.50	487.47	566.81 (10^6)
Aldehydes	234.07	267.39	300.72	334.04	367.37 (10^3)
Benzene	–	–	–	–	–
Tar	34.21	41.32	48.26	56.44	64.28 (10^6)
Volatile hydrocarbons	126.48	155.19	169.46	205.14	226.72 (10^6)
Formaldehyde	–	–	–	–	–
Organic acids	–	–	–	–	–
Methane	2.35	6.68	9.45	21.64	28.83 (10^6)
Hydrogen sulphide					
Total	–	–	1758.52	1758.52	3907.83
Metals					
Lead	–	–	–	1212.28	1377.86
Arsenic	–	–	1.95	1.95	4.34
Boron	–	–	13.68	13.68	30.39
Cadmium	–	–	–	–	–
Chromium	–	–	–	–	–
Mercury	–	–	7.82	7.82	17.37
Nickel	–	–	–	–	–
Zinc	–	–	–	–	–
Nitrogen oxides					
Total	58.53	80.70	107.21	139.09	169.89 (10^6)
Nitrous oxide	–	–	–	–	–
Sulphur oxides					
Total	4.40	8.75	12.59	24.85	30.52 (10^6)
SO_2	3353.89	4168.25	4954.64	5982.07	6970.70 (10^3)

Table 11.3 (continued)

	1985	1990	1995	2000	2005
Toxic hydrocarbons					
PAHs	–	–	–	–	–
Particulates					
Total	255.73	313.62	359.29	432.10	491.43 (10^6)
Size less than	–	–	–	189.59	215.49 (10^3)
Fugitive coal	–	–	–	–	–
Radioactive					
Carbon-14	–	–	–	–	– (Ci)
Iodine-131(elem)	–	–	–	–	– (Ci)
Iodine-131(non-el)	–	–	–	–	– (Ci)
Noble gases	–	–	–	–	– (Ci)
Radon	–	–	7.97	7.97	17.71 (Ci)
Tritium	–	–	–	–	– (Ci)
Ammonia					
Total	2661.70	3393.53	3527.85	3527.85	3547.25 (10^3)

(Source: Example of interlinked spreadsheet produced by LEAP)

Table 11.4 *Effluents to water courses resulting from a national energy demand scenario (Table 11.2) in kg*

	1985	1990	1995	2000	2005
Solids					
Total	12.17	15.52	16.05	16.06	16.06 (10^6)
Suspended	238.72	304.35	314.97	314.97	314.97 (10^3)
Dissolved	11.94	16.24	17.81	19.06	20.15 (10^6)
Oxygen demand					
Biochemical	683.14	874.80	909.09	913.74	917.85 (10^3)
Chemical	1129.03	1464.55	1540.25	1570.73	1597.62 (10^3)
Sulphates					
Total	6.83	162.22	318.24	504.65	669.13 (10^3)

Table 11.4 (continued)

	1985	1990	1995	2000	2005
Metals					
Total	2625.90	3616.81	4006.43	4332.99	4621.13
Cadmium	–	–	–	–	–
Chromium	–	201.70	406.28	651.20	867.30
Copper	–	–	–	–	–
Iron	–	–	–	–	–
Mercury	–	–	–	–	–
Zinc	–	201.70	406.28	651.20	867.30
Salts					
Total	–	–	–	–	–
Nitrates					
Total	–	–	–	–	–
Organic carbon					
Total	478.61	610.20	631.50	631.50	631.50 (10^3)
Oil and grease	71.62	167.95	248.88	341.95	424.07 (10^3)
Chlorides					
Total	2864.61	3844.29	4166.57	4399.78	4605.56 (10^3)
Ammonia					
Total	71.62	91.31	94.49	94.49	94.49 (10^3)
Phosphates					
Total	358.08	456.53	472.46	472.46	472.46
Cyanide					
Total	–	–	–	–	–
Radioactive					
Tritium	–	–	–	–	– (Ci)
Activation and FI	–	–	–	–	– (Ci)

(Source: Example of interlinked spreadsheet produced by LEAP)

Table 11.5 *Solid waste production from a national energy demand scenario (Table 11.2) in kg*

	1985	1990	1995	2000	2005
Mining waste					
Inert	–	–	–	–	–
Total	–	59.78	85.51	299.33	420.42 (10^6)
Ash					
Total	–	–	–	–	–
Scrubber sludge					
Total	–	–	–	–	–
Radioactive					
Low-level	–	–	–	–	– (Ci)
Low-level (vol)	–	–	–	–	– (Ci)

(Source: Example of interlinked spreadsheet produced by LEAP)

The development of a model is an iterative process, that is, the initial assumptions must be repeatedly improved upon as more information becomes available. This step is essential to obtain a reliable modelling tool. Eventually, the model results are translated into policy options about whose issues and practicalities judgement will need to be exercised. A useful integrated model enhances rather than dulls this judgement.

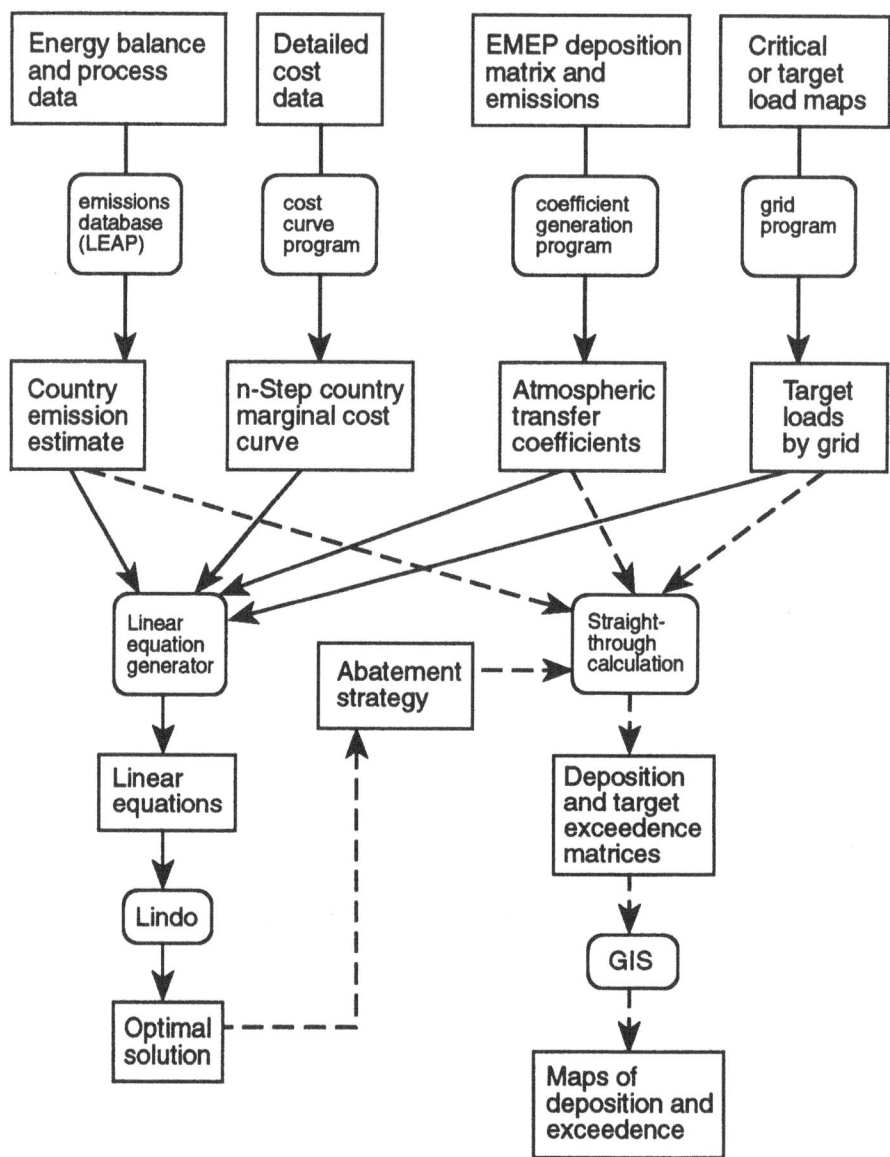

Figure 11.5 *The structure and data flows of a co-ordinated abatement strategy model*

(Source: SEI, 1991)

References

APDC (1985) *Integrated Energy Planning: A Manual.* Three volumes. Asian and Pacific Development Centre, Kuala Lumpur.

ASIAN DEVELOPMENT BANK (1992) *Integrated Energy-Environment Planning: Towards Developing a Framework.* Asian Development Bank, Manila.

BISWAS, A K, KHOSHOO, T N and KHOSLA, A (1990) *Environmental Modelling for Developing Countries.* Tycooly Publishing, London.

DANISH MINISTRY OF ENERGY (1990) *Energy 2000: A Plan of Action for Sustainable Development.* Danish Ministry of Energy, Copenhagen.

ELIASSEN, A, HOV, Ø, IVERSEN, T, SALTBONES, J and SIMPSON, D (1988) *Estimates of Airborne Transboundary Transport of Sulphur and Nitrogen.* EMEP/MSC-W Report 1/88. Norwegian Meteorological Institute, Oslo.

IAEA (1991) *Senior Expert Symposium on Electricity and the Environment: Key Issues Papers.* IAEA, Vienna.

MUNASINGHE, M (1990) *Energy Analysis and Policy.* Butterworth, London.

OECD (1991a) *The State of the Environment.* Organization for Economic Co-operation and Development, Paris.

OECD (1991b) *Environmental Indicators: A Preliminary Set.* Organization for Economic Co-operation and Development, Paris.

SEI (1991) *CASM: Co-ordinated Abatement Strategy Model.* SEI, Stockholm.

SEI-B (1992) *Long-range Energy Alternatives Planning System (LEAP): Overview.* SEI, Boston.

TUOVINEN, M (1992) 'Fluidized bed combustion of low-grade fuels'. *The Environmentally Sound Management of Low-grade Fuels.* SEI, Stockholm.

———————— Appendix A ————————
Environmental legislation for low-grade fuels

Approaches to the formulation and design of environmental legislation vary from country to country, but it is considered a fundamental requirement of sound environmental management (Estlander, 1990; Ottinger and Robinson, 1990; Rehbinder, 1990). This appendix provides an overview of some of the current methods for devising environmental legislation.

Environmental policy

Environmental legislation generally stems from a policy which, in a democratic society, is supposed to reflect the importance placed on environmental protection by that society. The development of an environmental policy may result from a combination of public pressure and government will and may be focused at a local, regional, national or international level. It may take the form of a specific policy statement, or it may become apparent over time through a build-up of regulatory mechanisms in different areas.

In many countries there is a ministry or department which is responsible for planning and overseeing environmental matters, but the onus for environmental protection will not rest solely with this agency. Legislation which will aid in environmental protection will usually be found in other ministries covering areas such as geology and mining, forestry, water resources, air quality, parks and nature conservation and agriculture. In some countries where there is no discrete government environmental agency, standards and regulations in these related areas will make up the only form of environmental legislation.

A policy of environmental protection will require an explicit statement of the goals and objectives behind the enactment and enforcement of legislation. It will also provide direction through definition of fundamental concepts referred to in the legislation such as what is meant by 'the environment'. In the past environment has often been synonymous with the physical environment (e.g. air, land and water). More recent definitions, such as that used by the Ontario Ministry of the Environment (1983) have taken a broader view and include:

- air, land or water;
- plant and animal life, including humans;
- the social, economic and cultural conditions that influence the lives of people or a community;
- any building, structure, machine or other device or thing made by people;
- any solid, liquid, gas, odour, heat, sound, vibration or radiation resulting directly or indirectly from human activities; or
- any part or combination of the foregoing and the interrelationships between any two or more of them.

Guidelines, regulations and standards

There are a number of control measures which may be taken to further the goal of environmental protection. These may be in the form of general guidance (guidelines) or more detailed technical requirements (regulations and standards) or both.

Environmental guidelines

Guidelines offer a more specific interpretation of the overall environmental policy. They will aid in the application of the environmental policy, generally giving a clearer indication of the intent of that policy. In many countries guidelines are not strictly enforceable by law, but by following a set of guidelines a project proponent, for example, an agency wishing to construct and operate an oil shale-processing facility, will be able to develop a proposal which is compatible with government policy and designed and constructed in a manner that implements the technical environmental regulations which are binding by law.

To ensure that environmental considerations are taken into account guidelines may be developed for undertakings such as the site selection of a power plant or the environmental impact assessment of a mine (Blaszczyk, 1992; Murphy, 1992). The development of guidelines requires a clear understanding of the guideline objectives (e.g. to ensure that protection of the environment is considered for all major undertakings; to ensure that members of the local community are consulted during site selection; to ensure that project development takes place in a manner which is conducive to meeting the prevailing regulations and standards; to maximize worker health and safety).

Once guidelines have been proposed, usually by a government agency, they will benefit from review and revision from experts in industry, academia, other government agencies, non-governmental organizations and from the public at large. Finalized guidelines should be made readily available to any party who wishes to use them. There is also a greater likelihood that guidelines will be followed if a representative of a government agency is given responsibility for aiding and reviewing work that is carried out according to the guidelines.

Regulations

Regulations, like guidelines, are derived from policy but are usually rules and restrictions that are actual laws with a penalty likely if violation occurs. Regulations include removal or control requirements for specific substances (e.g. permitted levels of SO_2 emissions by volume over a unit of time) and operating requirements (e.g. minimum allowable distance between the working face of an opencast mine and the nearest residence; requirement for the best available technology (BAT) at the time of installation or safe storage and treatment of solid waste from coal-fired power stations (Cope and Dacey, 1984). Table A1 provides an overview of the main control regulations in place in International Energy Agency (IEA) member countries for SO_2 emissions.

Standards

Many regulations are based on what are known as standards. Within an environmental context, standards are the maximum or minimum levels or requirements, often defined in a quantitative manner, to which an operation must conform. Standards are set for areas such as air quality (WHO, 1987), water quality (WHO, 1984) and worker health and safety (American Conference of Governmental Industrial Hygienists, 1988).

The development of standards requires a good deal of scientific expertise and financial resources and is a process which can take up to ten years (Ewers, 1991; Luken, 1990). For these reasons some countries feel that standard setting is beyond their means and will either set a limit or requirement for an operation on a case-by-case basis or will adopt standards suggested by an international agency such as the World Health Organization (WHO).

The setting of standards involves interaction with experts in the international community and between policy analysts and the government decision-making body and should include an opportunity for public review. It is also necessary to review all pertinent scientific information. Up-to-date information and reliable data come from what are known as criteria documents. Examples of sources of criteria documents include:

- WHO (1984 and 1987);
- American Conference of Governmental Industrial Hygienists (1988);
- US Environmental Protection Agency (EPA) (1986);
- Criteria Group of the Swedish National Board on Occupational Health and Safety;
- US National Institute of Occupational Safety and Health (1978);
- German MAK Commission (1988).

Table A1 *Types of SO_2 control regulations*

Country	Emissions standards	% Removal requirement	Coal sulphur limits	BAT requirements
Australia	*			
Austria	*		*	
Belgium	*			
Canada	*	(Note 2)		
Denmark	*		*	
European	*	*		*
Community				
(EC)	(Note 2)			
Finland	*		*	
France	(Note 2)			
Germany	*	*	*	*
Italy	*		*	
Japan	(Note 3)			
The Netherlands	*	*	*	
New Zealand			(Note 1)	*
Spain	*		(Note 4)	
Sweden	*			
Taiwan	*			
UK		*		*
USA	*	*		

1 Regional only
2 EC Directive on the limitation of emissions of pollutants into the air from large combustion plants, 1988
3 Set on plant-specific basis
4 Import coal only

(Source: Vernon, 1989)

There are three basic approaches to standard setting: technology-based standards, ambient-based standards and benefits-based standards.

Technology-based standards Technology-based standards are set with regard to the availability and cost of technology without taking into account the ambient environment (Luken, 1990). In this case limits on emissions or discharges would be tied to the levels currently achievable by existing technology.

Ambient-based standards Ambient-based standards are set with respect to known safety levels and do not involve a consideration of the cost or technological feasibility of achieving these levels (Luken, 1990). These are the standards most adamantly supported by those who advocate a safe and clean environment from the perspective of the health and welfare of the ecosystem.

Benefits-based standards Benefits-based standards are set following an assessment of the trade-offs between the risk to society from pollution of the environment and the costs (financial and otherwise) which society must bear to limit pollution (Luken, 1990).

Changing standards Standards continue to change as more information becomes available regarding the safe limits of exposure to or contamination by various substances and as environmentally cleaner technological processes evolve (e.g. Carter and Koksol, 1992; Shultz and Kitto, 1992). As Table A2 illustrates, this is often reflected in the difference in standards between new and existing plants.

Enforcement of regulations The enforcement of legislation necessitates monitoring levels of pollutant substances, reporting findings and prescribing penalties when regulations are not met. Monitoring requires personnel with relevant skills as well as monitoring equipment. Monitoring of emissions may take place by measuring the actual emissions or by monitoring the pollution control parameters (e.g. effectiveness of the ESP). Either of these indicators may be observed daily, although most monitoring is done on a longer-term basis (e.g. monthly), with samples taken at random intervals over a given period of time.

The results of monitoring are commonly given in monthly or yearly reports, although an accident or spill should be reported immediately. Failure to meet a given regulation may result in an appropriate authority meting out a punishment such as a monetary fine, a 'stop-work' order or, in extreme cases, imprisonment.

Table A2 *Selected international NO_x emissions limits*

Country	New Plants*		Existing Plants*	
	mg NO_2/m³	lb/10⁶ Btu	mg NO_2/m³	lb/10⁶ Btu
Austria	200–400	0.16–0.33	200–400	0.16–0.33
Belgium	200–800	0.16–0.65	–	–
Denmark	650**	0.53	**	
European Community (EC)	650–1300	0.53–1.60	–	–
Finland	200–400	0.16–0.33	400–620	0.33–0.50
Germany	200–500	0.16–0.41	200–1300	0.16–1.06
Italy	200–650	0.16–0.53	200–650	0.16–0.53
Japan	410–510	0.33–0.42	620–720	0.50–0.60
The Nether-lands	400–800	0.33–0.65	1100	0.90
Sweden	140	0.11	140–560	0.11–0.46
Switzerland	200–500	0.16–0.41	200–500	0.16–0.41
Taiwan	600–850	0.49–0.69	600–850	0.49–0.69
UK	650	0.53	–	–
USA	615–980	0.500.80	553–614	0.450.50***

*   Conversion Factors: 8.14 x 10⁻⁴ lb/10⁶ Btu per mg/nm³
                                            350 m³ flue gas/GJ fuel input
                                            30 GJ/t
**   In addition to 'bubble' principle for utilities
*** Dry bottom wall-fired and tangential-fired only; other limits pending

(Source: Hjarlmarsson and Soud, 1990)

Low-grade fuel regulations

There are very few regulations which specifically mention low-grade fuels, but they are covered by regulations in use in many countries pertaining to solid fuels (Estlander, 1990). In many European countries emissions from MSW combustion are strictly regulated. EPA also has special emission monitoring programmes for MSW. Regulations for open pit mining will often apply to peat, lignite and oil shale. In Finland, local authorities can impose conditions on peat harvesting over 50 hectares (Oikarinen, 1986). Combustion of solid fuels for energy production is the most heavily regulated stage of the fuel cycle. These regulations often focus on individual pollutants such as SO_2, NO_x, CO, volatile organic compounds and halogens. Emission limits set for energy production are often the same for all solid fuels. This means that low-grade fuels have to achieve the same emission standards as higher-quality fuels with fewer impurities. The lower cost of low-grade fuels, in some cases, may help to offset the cost of better pollution control.

In many countries the solid waste produced during processing or combustion of low-grade fuels is also covered under regulations pertaining to the containment of heavy metals and organic compounds. Legislation relating to the residue from the combustion of MSW and industrial sewage sludge is currently being developed by the EPA in the United States due to the known high levels of trace elements and organic compounds in these fuels (Vancil, Parrish and Palazzolo, 1992).

References

AMERICAN CONFERENCE OF GOVERNMENTAL INDUSTRIAL HYGIENISTS (1988) *Documentation of the Threshold Limit Values for Chemical Substances and Physical Agents in the Work Environment.* 4th Edition. ACGIH, Cincinnati.

BLASZCZYK, B (1992) *Legal Regulations and Environmental Policy.* Ministry of Environmental Protection, Poland (unpublished report).

CARTER, H R and KOKSAL, C G (1992) *On-line Imaging and Emissivity Measurements to Determine Furnace Cleanliness.* Diamond Power Speciality Company, Lancaster.

COPE, D R and DACEY, P W (1984) *Solid Residues from Coal Use Disposal and Utilization.* IEA Coal Research, London.

ESTLANDER, A (1990) 'Review of environmental regulations'. In *VTT Symposium 108* (ed by M Korhonen). Technical Research Centre of Finland, Espoo.

EWERS, U (1991) 'Standards, guidelines and legislative regulations concerning metals and their compounds'. *Metals and Their Compounds in the Environment.* VCH, Wernheim.

HJARLMARSSON, A-K and SOUD, H N (1990) *Systems for Controlling NO$_x$ from Coal Combustion.* IEA Coal Research, London.

LUKEN, R A (1990) *Efficiency in Environmental Regulation: A Benefit-Cost Analysis of Alternative Approaches.* Kluwer Academic Publishers, Boston.

MAK (1988) *Maximum Concentrations at the Workplace and Biological Tolerance Values for Working Materials.* VCH Verlagsgesellschaft, Weinheim.

MURPHY, I L (1992) *Environmental Compliance: Matching Policy Strategies with Optimal Coal Utilization.* Regional Environmental Center, Budapest (unpublished report).

OIKARINEN, S (1986) 'Environment regulations for peat use in Finland'. *Bio Energi 86.* Goteborg (in Swedish).

ONTARIO MINISTRY OF THE ENVIRONMENT (1983) *General Guidelines for the Preparation of Environmental Assessments.* Ontario Ministry of the Environment, Toronto.

OTTINGER, R and ROBINSON, N (1990) 'The use of law to incorporate the cost of protecting the environment into the pricing of electricity'. *Report of the Advanced Seminar on Economic and Legal Aspects of Pollution Abatement Strategies in Europe.* UNESCO/ROSTE, Venice.

REHBINDER, E (1990) 'Ambient and emissions oriented strategies of air pollution control: a comparison of their advantages and disadvantages and their relationships to general principles of environmental law'. *Report of the Advanced Seminar on Economic and Legal Aspects of Pollution Abatement Strategies in Europe.* UNESCO/ROSTE, Venice.

SHULTZ, S and KITTO, J B (1992) *Steam: Its Generation and Use.* Babcock and Wilcox, New Orleans.

US ENVIRONMENTAL PROTECTION AGENCY (1986) *Air Quality Criteria for Lead.* Vols. 1-4. Environmental Criteria and Assessment Office, North Carolina.

US NATIONAL INSTITUTE FOR OCCUPATIONAL HEALTH AND SAFETY (1978) *Criteria for a Recommended Standard.* US Department of Health, Education and Welfare, Washington.

VANCIL, M A, PARRISH, C R and PALAZZOLO, M A (1992) *Emissions of Metals and Organics from Municipal Wastewater Sludge Incinerators.* EPA, Cincinnati.

VERNON, J L (1989) *Market Impacts of Sulphur Control: The Consequences for Coal.* IEA Coal Research, London.

WHO (1987) *Air Quality Guidelines for Europe.* WHO Regional Publications, Copenhagen.

WHO (1984) *Guidelines for Drinking Water Quality.* Vols. 1-3. WHO, Geneva.

Glossary

acid mine drainage
Coal contains iron pyrite (FeS_2) which when dissolved in water lowers the pH values of surrounding water and soils, thus creating more acidic conditions. For this reason water seeping through coal-mining sites is known as acid mine drainage and requires collection and control if pollution of associated aquatic ecosystems is to be avoided.

agro-residues
Also known as crop residues, these include what are often considered as waste products from agricultural production such as leaves and shafts of plants and nut shells.

anaerobic digestion
A biological process where, under oxygen-free conditions, organic matter breaks down into carbon dioxide and methane.

anthracite
A high calorific value coal with a high carbon content. Anthracite burns with little smoke and is used in domestic cooking and heating.

'as received'
The condition of mined material prior to processing.

ash
Minerals and other inorganic matter which are chemically or otherwise closely associated with a fuel, and which will remain following complete combustion of the fuel.

ash content
The percentage of ash, by weight, in a given fuel sample, determined under specific parameters of pressure and combustion temperature in a laboratory setting.

ash-free
Indicates that any information being given (e.g. percentage of sulphur) has been calculated for a fuel sample from which the ash content has been subtracted.

baghouses
See fabric filtration.

beneficiation
The processing and upgrading of a fuel to obtain desired characteristics such as a higher calorific value by weight.

biogas
A product of organic matter decaying under oxygen-free conditions, primarily composed of methane and carbon dioxide.

biomass
A variety of organic substances such as wood crop residues, dung, and dried water plants.

bituminous coal
A black coal, generally falling between the lower calorific value sub-bituminous coal and the higher calorific value anthracite.

boiler
A container in which fuel is burned to produce heat used to produce steam that may be used for space heating or driving a turbine for the generation of electricity.

briquetting
The compaction of fuel dust or small fragments of fuel, with or without an adhesive or binding material, to produce pellets, bricks or ball-shaped pieces of fuel.

brown coal
See lignite.

calorific value
The heat given off during combustion of a standard quantity of fuel at a standard temperature.

charcoal
Carbonized wood achieved by burning wood under near oxygen-free conditions.

chemical feedstocks
Specific chemicals derived from more complex substances for purposes of commercial production of chemicals.

clarifier sludge
A slurry or mud-like substance which is a by-product of removing heavier and more opaque substances from a liquid.

co-generation
Producing energy which meets two purposes at the same time, for example, producing steam for space heating and using some of the steam to generate electricity for the plant.

coalification
The transformation of plant material over time, under a variety of physical and chemical conditions such as high pressure and high temperature, from oxygen-rich plant debris to carbon-rich, oxygen-poor coal. The degree of coalification generally, but not always, relates to age, with the youngest coal being lignite and the oldest coal, anthracite.

coking
Carbonization of coal under oxygen-poor conditions resulting in semicoke or coke, a coal product which no longer contains volatiles.

combustion
Burning of a substance in the presence of oxygen during which heat and light are released.

dioxins
A general term for aromatic or polyaromatic polyhalogenated compounds which can result from the burning of chlorinated compounds such as plastics. Some dioxins are highly toxic (e.g. tetrachlorodibenzo-*p*-dioxin).

domestic energy
Energy used in the home for daily living requirements such as cooking, light and heat.

effluent
Waste resulting from a production system in liquid, gaseous or solid form. If the effluent is contaminated, its release into the environment must be prevented.

electrostatic precipitator
Used for particulate removal from flue gas, electrostatic precipitators (ESPs) comprise two plates, 15 to 25 cm apart, between which are wire electrodes. When ionized gas is passed between the plates, dust particles attach to the electrodes emitting the opposite charge.

emissions
Substances discharged into the environment.

end-of-pipe solutions
This term is used to refer to environmental protection procedures that focus only on controlling, reducing or eliminating the final output of a system or process (e.g. pollutant emissions) without taking into consideration the need for or benefits of redesigning or restructuring the existing process.

energy efficiency
Releasing and applying energy from a fuel with the least possible waste of resources.

environmental impacts
The effects of a project or undertaking on the physical, man-made and social environments (e.g. air pollution, job creation, loss of arable land).

external costs
Costs that are traditionally not calculated when determining the costs of producing specific goods. Also referred to as environmental costs, these could include waste management, cost of preventing or compensation for disruption to air and water quality or land use and the loss of a potential resource to future generations.

fabric filtration
Passage of flue gas through dense cloth filters made of synthetic material such as nylon, Dacron or woven fibreglass to remove dust and fly ash. The fabric filters are also known as baghouses.

fines
Very fine particles of fuel or very small fragments of fuel.

fixed bed combustion
The burning of solid fuel which is sitting on an open grid or grate that does not move or, if it does, moves only to remove a build-up of ash.

flash drying
Recycling of hot flue gases to dry lignite.

flue gas cleaning
General term given to various processes used to remove particulates and other unwanted substances from waste gases emitted from a production process.

fluidized bed combustion
Burning of fuel on a bed of turbulent air rather than on a grate which does not move.

fossil fuel
Fuels comprising fossilized organic materials.

fuel cycle
A term used to describe the steps involved in obtaining and using a fuel. The major stages of the cycle include mining, processing, transport, conversion and use.

furans
Aromatic or polyaromatic polyhalogenated compounds such as polychlorinated dibenzofurans.

gasification
The conversion of a solid or liquid to a gaseous state.

grade
For energy purposes grade is used to identify a fuel according to its value for a given end use. Factors which can contribute to grade include calorific value, ash, moisture, oxygen and sulphur content.

halogens
Substances generally found in a highly reactive gaseous state including chlorine, fluorine, bromine and iodine and their associated compounds such as aromatic or polyaromatic polyhalogenated compounds (e.g. dioxins and furans).

hemic peat
Peat of a medium decomposition with 40–70 per cent fibre – the best quality peat for fuel purposes.

holding/settling ponds
Areas where liquid effluent is contained, at least until the heavier particles contained in the liquid have settled. This residue may be taken to landfill, recycled or treated depending upon its composition.

hydrolysis
Decomposition of a substance by the chemical action of water.

hydropyrolysis
The conversion of a liquid product from a solid fuel during which the solid fuel undergoes chemical destruction at 480–490°C under hydrogen pressure.

impurities
General term for undesired substances with respect to the end use of a fuel.

indicators of environmental quality
Measures for determining the well-being of the environment.

industrial waste
Waste generated by industrial processes (e.g. equipment manufacture, oil refining, chemical production).

inorganic matter
Mineral component of a fuel which becomes the ash which remains following combustion.

integrated energy planning
Energy planning that goes beyond simple demand calculations to take into account a number of interrelated features such as the most efficient and valuable use of energy resources along with enhancement of the welfare and quality of life of a population.

integrated gasification combined cycle
The gasification of fuel as part of a larger system of energy production.

intellectual property
Ideas and information such as plans, patents, designs, trademarks, copyrights and traditional indigenous knowledge, the rights to which are sometimes bought or sold.

internal costs
Also known as industry costs, these are the costs involved in carrying out a given production process including, among others, costs for equipment, land acquisition and employees.

kerogen
Carbon-rich material contained in oil shale.

lignite
A brown coal which is generally the least dense and least decomposed of the various ranks of coal. Also referred to as brown coal, particularly in Eastern Europe.

liquefaction
Conversion of a solid to a liquid.

low-grade fuel
Defined here as a fuel with a calorific value of less than 17.5 MJ/kg (4182 kcal/kg) or a fuel with a calorific value greater than 17.5 MJ/kg and an impurity content with a potentially significant environmental impact such as a sulphur content greater than 3 per cent or an ash content greater than 20 per cent.

maceral group
The characteristics of the plant material from which coal is formed.

methanol
Methyl alcohol which may be produced from wood or peat and which can be used as a liquid fuel and is also a commonly used solvent.

milled peat
Shredded peat, usually achieved by machine-cutting peat where it is found, then drying the cut peat in the open air.

mining spoil
The mined material left behind at the mining site following the removal of the desired fuel.

MSW
See municipal solid waste.

municipal solid waste
Solid household and commercial waste is commonly referred to as municipal solid waste or MSW since it is often the responsibility of the local government authority to provide for its disposal.

oil shale
Fossil fuel formed primarily underwater from small sea plants and animals and found in alternating layers with limestone. The percentage of fuel, by weight, is relatively low and therefore produces a very high percentage of waste (spent shale) for every unit of fuel extracted.

olefin
A hydrocarbon from which many substances are derived, including ethylene, propylene and butadiene, that are used to manufacture synthetic rubber, nitroglycerin, printing inks and carpet backing.

opencast mining
The mining of a mineral deposit by removing the overlying ground cover.

organic material
Complex carbon compounds, the majority of which are produced by plants and animals.

PAHs
See polycyclic aromatic hydrocarbons.

particulates
Particles formed during combustion including dust and fly ash which may range in size from fairly large visible particles to submicron particles.

peat
The youngest of the fossil fuels, peat is comprised of layers of plant material built up over time under waterlogged conditions.

pellet
See briquetting.

phenols
Carboxylic acid, a toxic substance, which can be extracted from fossil fuels and which is used in the production of adhesives.

polycyclic aromatic hydrocarbons
A major group of carbon compounds formed during fuel combustion, including benzo(a)pyrene (BaP) and dibenz(a,h)anthracene (DBahA),which can pose a health hazard to humans and other animals.

proximate analysis
Test to determine the volatile matter of a fuel, the moisture present at 105°C, the ash residue and the fixed carbon of a coal sample.

pulverized fuel
Fuel that is broken down physically into smaller fragments, even to powder form, to create greater surface area for reactions.

pyrolysis
Decomposition of a substance by the application of heat.

rank
For energy purposes rank is used most commonly to refer to coal. Low-rank coal is distinguished from higher-ranked coals by a high moisture content, low heating value, high oxygen content and generally alkaline but highly variable inorganic content. Two tests are used to determine coal rank: proximate analysis, standard tests to determine the volatile matter, the equilibrium moisture present at 105°C, the ash residue and the fixed carbon; and ultimate analysis, which provides information on the carbon, hydrogen, sulphur and nitrogen present in dry, ash-free coal.

RDF
See refuse-derived fuel.

reclamation
Refers to the restoration of land following opencast mining, often replacing subsoil, topsoil and overburden.

refuse-derived fuel
Processed municipal solid waste.

regenerable processes
Processes of flue gas desulphurization in which sulphur is absorbed chemically or thermally and a saleable product (e.g. liquified sulphur dioxide, elemental sulphur or sulphuric acid) is produced.

renewable fuel
Fuel which can replenish itself such as solar energy, tidal energy and geothermal energy. Biomass (indirectly solar energy) is also considered a renewable fuel.

reserves
The known quantities of fuel which are obtainable using current methods.

resources
All known and unknown quantities of a fuel, whether recoverable or not.

retorting
The application of heat in gas or steam form to carbonize lignite or coal or to liberate shale oil from oil shale. It can take place in a special device known as a retort (surface retorting) or more commonly in the case of oil shale in-ground where the deposit is found (in situ retorting).

scrubbers
The various materials added to flue gas to remove unwanted sulphur dioxide (e.g. limestone).

SNG
See synthetic natural gas.

social impacts
Effects on lifestyle, community and culture brought about by a project.

spent shale
The shale which is left after the removal of oil from oil shale.

standards
Within an environmental context, standards are the maximum or minimum levels or requirements, often defined in a quantitative manner, to which an operation must conform.

synthetic natural gas
Gas which results from a process designed to gasify a fuel.

trace elements
Heavy metals such as arsenic, cadmium, copper, iron, lead, mercury and vanadium which are usually measured in micrograms and comprise a very small percentage of fuel by weight or volume.

ultimate analysis
Tests conducted to determine the carbon, hydrogen, sulphur and nitrogen present in dry, ash-free coal.

upgrading
Processes carried out to enhance the properties of a fuel for a given end use.

volatile matter
Matter which evaporates during the combustion of a fuel.

wastewater
Liquid waste resulting from a production process.

Index